LABORATORY AND CLINICAL RESEARCH

BIOLOGICAL LABORATORIES

SAFETY REPORTING AND OVERSIGHT CONSIDERATIONS

LABORATORY AND CLINICAL RESEARCH

Additional books in this series can be found on Nova's website under the Series tab.

Additional e-books in this series can be found on Nova's website under the e-books tab.

SAFETY AND RISK IN SOCIETY

Additional books in this series can be found on Nova's website under the Series tab.

Additional e-books in this series can be found on Nova's website under the e-books tab.

LABORATORY AND CLINICAL RESEARCH

BIOLOGICAL LABORATORIES

SAFETY REPORTING AND OVERSIGHT CONSIDERATIONS

JENSEN HOPKINS
AND
NEAL CALDWELL
EDITORS

NOVA BIOMEDICAL
New York

For permission to use material from this book please contact us:
Telephone 631-231-7269; Fax 631-231-8175
Web Site: http://www.novapublishers.com

LIBRARY OF CONGRESS CATALOGING-IN-PUBLICATION DATA

ISBN: 978-1-62257-822-1

Published by Nova Science Publishers, Inc. † New York

CONTENTS

PREFACE

As the number of biological labs increases, so do the safety risks for lab workers. Data on these risks--collected through a safety reporting system (SRS) from reports of hazards, incidents, and accidents--can support safety efforts. However, no such system exists for all biological labs, and a limited system, managed by the Centers for Disease Control and Prevention (CDC) and the Animal and Plant Health Inspection Service (APHIS), applies to only a subset of these labs. Given the increase in biological labs and therefore risks, it is essential to understand the sources of risk and how to communicate them. This book examines the collection of data on accidents, incidents, and hazards in biological labs, that can help identify accident precursors. If the precursors can be identified, communicated, and eliminated, the occurrence of accidents, in particular those resulting in injury or death, might be prevented.

Chapter 1 - As the number of biological labs increases, so too do the safety risks for lab workers. Data on these risks—collected through a safety reporting system (SRS) from reports of hazards, incidents, and accidents—can support safety efforts. However, no such system exists for all biological labs, and a limited system—managed by the Centers for Disease Control and Prevention (CDC) and the Animal and Plant Health Inspection Service (APHIS)—applies to only a subset of these labs. While a national SRS has been proposed, design and implementation are complex. In this context, GAO was asked to identify lessons from (1) the literature and (2) case studies; and to apply those lessons to (3) assess CDC and APHIS's theft, loss, or release (TLR) system for select agents, such as anthrax, and (4) suggest design and implementation considerations for a labwide SRS. To do its work, GAO analyzed SRS literature; conducted case studies of SRSs in aviation, commercial nuclear, and health care industries; and interviewed agency officials and biosafety specialists.

Chapter 2 - The federal government responded to the September 11, 2001, terrorist attacks and the subsequent anthrax attacks with increased focus on and funding for biodefense. A key consideration in this response was addressing shortages in diagnostic, clinical, and research laboratory capacity. Several departments and agencies have increased or are in the process of increasing their laboratory capacity. High-containment laboratories play a critical role in the biodefense effort, offering the hope of better responses to an attack and a better understanding of the threat posed by bioterrorism. However, they also could increase the risk of a biological attack by serving as a potential source of materials or training. Indeed, the Commission on the Prevention of Weapons of Mass Destruction Proliferation and Terrorism recommends tightening government oversight of high-containment laboratories. Policymakers have become increasingly interested in the oversight of these facilities following reports of accidents, regulatory noncompliance, and community resistance. The increase in high-containment laboratory capacity has raised new policy questions and emphasized existing ones. How much laboratory capacity is enough? What is the necessary federal investment? Should laboratories be consolidated or dispersed? What plans exist to coordinate multiple agency efforts to expand high-containment laboratory capacity? Does increasing laboratory capacity increase the risk of accidents and the opportunity for purposeful misuse? What is an acceptable balance between the benefits these laboratories provide and the risks they pose? Interested Members of Congress might take action to address some or all of these concerns. Alternatively, they might defer action until efforts currently under way assess and make recommendations regarding the existing regulatory structure. If Congress chooses to enhance oversight, it might require a survey of existing facilities and their use and a national needs assessment, perhaps barring further construction until these are complete. Stakeholders could focus on enhancing self-regulatory activities such as improving or standardizing laboratory worker training or building a mechanism for sharing lessons learned. Rather than relying on self- regulation, policymakers might enhance oversight through additional regulation of high-containment facilities, requiring laboratory or personnel certification, or by broadening the Select Agent Program. Which agencies should implement any new mandates remains an open question. Biocontainment technologies are widely used by scientists around the world. Efforts to increase control of U.S. high-containment laboratories may put domestic industry at a competitive disadvantage and inhibit international academic collaboration. Absent international harmonization, the United States can only partially address the threat of a high-containment laboratory being the source of a bioterror weapon. A key task for policymakers is to define their goals for enhancing oversight of high-containment laboratories. The focus of the oversight effort may affect which policy issues are addressed. For example, focusing on a registry of existing high-containment laboratory capacity may improve planning, coordination, and efficiency of use but provide relatively limited security benefits. Similarly, a rigorous oversight program including facility and personnel licensure, mandatory training, and restricted construction of new facilities may provide security benefits at the cost of regulatory burden, increased federal expenditures, and impeded scientific progress in countermeasure research, bioforensics, and public health. When weighing options to address these complex policy issues, policymakers may have to reconcile many competing and potentially conflicting national needs.

In: Biological Laboratories
Editors: J. Hopkins and N. Caldwell
ISBN: 978-1-62257-822-1

Chapter 1

BIOLOGICAL LABORATORIES: DESIGN AND IMPLEMENTATION CONSIDERATIONS FOR SAFETY REPORTING SYSTEMS*

The United States Government Accountability Office

ABBREVIATIONS

ABSA	American Biological Safety Association
AHRQ	Agency for Healthcare Research and Quality
APHIS	Animal and Plant Health Inspection Service
ASAP	Aviation Safety Action Program
ASM	American Society for Microbiology
ASRS	Aviation Safety Reporting System
BMBL	Biosafety in Microbiological and Biomedical Laboratories
BSC	biological safety cabinet
BSL	biosafety level
CDC	Centers for Disease Control and Prevention
ERT	event review team
FAA	Federal Aviation Administration
FOIA	Freedom of Information Act

* This is an edited, reformatted and augmented version of The United States Government Accountability Office publication, Report to Congressional Requesters GAO-10-850, dated September 2010.

HHS	Department of Health and Human Services
INPO®	Institute of Nuclear Power Operators
LAI	laboratory-acquired infection
MMWR	Morbidity and Mortality Weekly Report
NAPA	National Academy of Public Administration
NASA	National Aeronautics and Space Administration
NCPS	National Center for Patient Safety
NIH	National Institutes of Health
NN®	Nuclear Network
NRC	Nuclear Regulatory Commission
NTSB	National Transportation Safety Board
OIG	Office of Inspector General
OSHA	Occupational Safety and Health Administration
PSIS	Patient Safety Information System
PSRS	Patient Safety Reporting System
SAP	Select Agent Program
SEE-IN®	Significant Event Evaluation—Information Network
SRS	safety reporting system
TMI	Three Mile Island
TLR	theft, loss, release
USDA	Department of Agriculture
VA	Department of Veterans Affairs
VDRP	Voluntary Disclosure Reporting Program
VSP	Voluntary Safety Programs Branch

WHY GAO DID THIS STUDY

As the number of biological labs increases, so too do the safety risks for lab workers. Data on these risks—collected through a safety reporting system (SRS) from reports of hazards, incidents, and accidents—can support safety efforts. However, no such system exists for all biological labs, and a limited system—managed by the Centers for Disease Control and Prevention (CDC) and the Animal and Plant Health Inspection Service (APHIS)—applies to only a subset of these labs. While a national SRS has been proposed, design and implementation are complex. In this context, GAO was asked to identify lessons from (1) the literature and (2) case studies; and to apply those lessons to (3) assess CDC and APHIS's theft, loss, or release (TLR) system for select agents, such as anthrax, and (4) suggest design and implementation considerations for a labwide SRS. To do its work, GAO analyzed SRS literature; conducted case studies of SRSs in aviation, commercial nuclear, and health care industries; and interviewed agency officials and biosafety specialists.

WHAT GAO RECOMMENDS

GAO recommends that, in developing legislation for a national SRS for biological labs, Congress consider provisions for certain system features. GAO also recommends three improvements to the CDC and APHIS TLR system.

HHS disagreed with the first two recommendations and partially agreed with the third. USDA agreed with the three recommendations.

WHAT GAO FOUND

According to the literature, effective design and implementation of a safety reporting system (SRS) includes consideration of program goals and organizational culture to guide decisions in three key areas: (1) reporting and analysis, (2) reporter protection and incentives, and (3) feedback mechanisms. Program goals are best identified through stakeholder involvement and organizational culture, through assessment.

Case studies of SRSs in three industries—aviation, commercial nuclear, and health care—indicate that (1) assessment, dedicated resources, and management focus are needed to understand and improve safety culture; (2) broad reporting thresholds, experience-driven classification schemes, and local-level processing are useful SRS features in industries new to safety reporting; (3) strong legal protections and incentives encourage reporting and prevent potential confidentiality breaches; and (4) a central, industry-level unit facilitates lesson sharing and evaluation.

While the CDC and APHIS Select Agent Program (SAP) has taken steps in the three key areas to improve the usefulness of the TLR system for select agents, steps for improvement remain. Specifically, the agencies have taken steps to better define reportable events, ensure the confidentiality of reports, and dedicate resources to use TLR data for safety improvement. However, lessons from the literature and case studies suggest additional steps in the three key areas to enhance the usefulness of the system. For example, lowering reporting thresholds could provide precursor data and limited immunity could increase the incentive to report. Finally, the CDC and APHIS are in a unique position—as recognized authorities in the lab community and with access to TLR reports from across the industry—to guide SRS evaluation and ensure safety lessons are broadly disseminated.

For a national safety reporting system for all biological labs, existing information—about labs' organizational culture and the lab community's limited experience with SRSs—suggests the following features in the three key areas:

- *Reporting and analysis.* Reporting should be voluntary; available to all workers; cover hazards, incidents, and less serious accidents; accessible in various modes (Web and postal); and with formats that allow workers to report events in their own words to either an internal or external SRS system.
- *Reporter protections and incentives.* Strong confidentiality protections, data deidentification processes, and other reporting incentives are needed to foster trust in reporting.
- *Feedback mechanisms.* SRS data should be used at both the local and industry levels for safety improvement. An industry-level entity is needed to disseminate SRS data and to support evaluation.

September 10, 2010

The Honorable Joe Barton
Ranking Member
Committee on Energy and Commerce
House of Representatives

The Honorable Michael Burgess
Ranking Member
Subcommittee on Oversight and Investigations
Committee on Energy and Commerce
House of Representatives

The Honorable Greg Walden
House of Representatives

The growing federal emphasis on identifying and protecting against biological weapons attacks, as well as other factors, have led to an increase in the number of biological laboratories in the United States. Although data suggest that injury and illness rates for these labs are below that of general industry, working with infectious agents always involves inherent risk.[1] To date, catastrophes have been avoided in the United States, although serious injuries and deaths have occurred among laboratory workers.[2] These injuries and deaths might have been prevented had relevant data on safety been quickly shared throughout the laboratory community. For example, two microbiologists died in July and December 2000 of a laboratory-acquired infection (LAI) from exposure to bacterium *Neisseria meningitidis.* In a review of how often this LAI had occurred, investigators found 14 previously unreported LAIs from exposure to the bacteria—8 of which were fatal.[3] Had these LAIs been reported and the safety issues surrounding this specific bacterium communicated earlier, the two deaths in 2000 might have been prevented.

Given the increase in biological labs and therefore risks, it is essential to understand the sources of risk and how to communicate them.[4] These sources can best be identified through the collection of safety data. Such data can come from accidents that result in injuries or deaths. However, they can also come from concerns about hazardous conditions or incidents such as errors without consequences, near misses, or close calls. Collecting data on accidents, incidents, and hazards can help identify accident precursors—the actions, nonactions, processes, and environmental or mechanical conditions that can lead to accidents.[5] If the precursors can be identified, communicated, and eliminated, the occurrence of accidents—in particular those resulting in injury or death— might be prevented. Safety reporting systems (SRS) are a key tool industries use to collect such information. However, there is no national labwide SRS for quickly and efficiently collecting, analyzing, and communicating such information for biological labs. Nevertheless, some mechanisms exist through which such data might be communicated. For example, incidents of LAIs are sometimes reported in academic journals, in the U.S. Department of Health and Human Services' (HHS) Centers for Disease Control and Prevention's (CDC) *Morbidity and Mortality Weekly Reports* (MMWR), or as a result of Occupational Health and Safety Administration (OSHA) regulations. However, there are a variety of barriers to reporting through these mechanisms, and it is generally acknowledged that LAIs are underreported because of concerns about stigma or punishment. Consequently, a great deal of potential safety data is never communicated.

In addition, the CDC and the U.S. Department of Agriculture's (USDA) Animal and Plant Health Inspection Service (APHIS) together maintain a mandatory reporting system for theft, loss, and release (TLR) of select agents,[6] as required under the select agent regulations.[7] However, we have found lapses in labs reporting to this program,[8] suggesting the need for improvement. Moreover, the Select Agent Program regulates only those labs that possess, use, and transfer select agents and toxins, and therefore covers only a fraction of U.S. biological labs for which there is no SRS.[9] Consequently, a great deal of valuable safety data falls through the cracks, and potentially avoidable accidents continue to occur.

Recognizing the need for an effective mechanism to collect safety data, bills were introduced in both the Senate and House of Representatives that, if enacted, would establish a new SRS for all biological labs.[10] While this legislation provides a framework for establishing such a system, questions remain about what constitutes the most effective design and implementation features for a biological lab SRS. Despite these questions, it is known that effective design and implementation include the use of existing information, such as from the literature and case studies, to identify lessons learned that can guide decisions. For example, when the health care industry began to explore the potential of SRSs for hospitals, many in the industry looked to the literature and other industries, such as aviation, to identify lessons learned for design and implementation. Similarly, for biological labs, although they are a unique working environment, information from the literature and other industries can identify lessons learned for the design and implementation of a lab SRS.[11] You therefore asked us to identify lessons for designing and implementing an effective lab safety reporting system, from (1) the literature and (2) case studies of SRSs in the airline, commercial nuclear power, and health care industries; and to apply those lessons to (3) assess the theft, loss, and release reporting system, part of the Select Agent Program, and (4) suggest design and implementation considerations for a national safety reporting system for all biological labs.

To accomplish our objectives, we (1) reviewed an extensive selection of both academic and applied literature related to safety science (organizational safety and human factors) and SRS evaluation across a wide variety of industries; (2) conducted case studies of SRSs in the aviation, commercial nuclear power, and health care industries by reviewing relevant documentation and academic literature, observing safety task force and reporting system committee meetings, and conducting open and structured interviews of agency officials, as well as SRS and human factors experts in the three industries; (3)

interviewed national and international biosafety specialists, relevant HHS and USDA officials, biological laboratory directors, and biosafety officers; and (4) applied criteria—derived from our review of the literature and case studies—for improving the Select Agent Program reporting system and for designing and implementing an SRS for all biological labs. With respect to the case studies, while we collected information on a wide variety of safety reporting programs in the three industries—and in some cases comment on these different programs—we primarily developed our lessons from one reporting program in each of the three industries. Specifically, we developed lessons from the Federal Aviation Administration's (FAA) National Aeronautics and Space Administration (NASA)-run Aviation Safety Reporting System (ASRS) in aviation; the Institute of Nuclear Power Operation's (INPO®) Significant Event and Evaluation Information Network (SEE-IN®) system in commercial nuclear power; and the Department of Veterans Affairs' (VA) health care reporting program, which includes the Patient Safety Information System (PSIS) and the Patient Safety Reporting System (PSRS). We chose to focus on these programs because they represent fairly long-standing, nonregulatory, domestic, industrywide, or servicewide reporting programs. For more detailed information on our methods, please see appendix I.

We conducted this performance audit from March 2008 to September 2010 in accordance with generally accepted government auditing standards. Those standards require that we plan and perform the audit to obtain sufficient, appropriate evidence to provide a reasonable basis for our findings and conclusions based on our audit objectives. We believe that the evidence obtained provides a reasonable basis for our findings and conclusions based on our audit objectives.

BACKGROUND

Biocontainment laboratories—designed with specific environmental, storage, and equipment configurations—support containment efforts in the day-to-day work with biological agents. These labs are designed, constructed, and operated to (1) prevent accidental release of infectious or hazardous agents within the laboratory and (2) protect lab workers and the environment external to the lab, including the community, from exposure to the agents. For example, the biological safety cabinet (BSC) is laboratory safety equipment that is used when manipulating infectious organisms. BSCs are enclosed cabinets with mechanisms for pulling air away from the worker and into a HEPA filter, which provides protection for the worker and prevents releases into the environment. BSCs might be designed with a limited workspace opening, or they might be completely enclosed with only gloved access and air pressure indicators to alert users to potential microbial releases. The selection of the BSC would depend on the (1) lab's risk assessment for the specific agent and (2) nature of work being conducted, as guided by the Biosafety in Microbiological and Biomedical Laboratories (BMBL), and other relevant guidance, such as OSHA regulations and National Institutes of Health (NIH) guidelines for research involving recombinant DNA.

There are four biosafety levels (BSL). These levels—consisting of a combination of laboratory practices, safety equipment, and laboratory facilities—are based on the type of work performed, information about the infectious agent, and the function of the laboratory. These levels include combinations of laboratory practices and techniques, safety equipment, and facilities that are recommended for labs that conduct research on infectious mircro-organisms and toxins:

- *Biosafety level 1 (BSL-1)* is suitable for work with agents not known to consistently cause disease in healthy adults and present minimal potential hazard to laboratory personnel and the environment.
- *Biosafety level 2 (BSL-2)* is suitable for work with agents that pose moderate risks to personnel and the environment.
- *Biosafety level 3 (BSL-3)* is suitable for work with indigenous or exotic agents that may cause serious and potentially lethal disease, if inhaled.
- *Biosafety level 4 (BSL-4)* is required for work with dangerous and exotic agents that pose a high risk of life-threatening disease or have aerosol or unknown transmission risk.

Examples of agents and toxins used within these labs include those that primarily affect

- humans and animals, such as *Botulinum neurotoxin*, a naturally occurring poison, lethal to humans and animals, but used for medical and cosmetic purposes in drugs such as Botox;
- animals, such as foot-and-mouth disease (FMD), a highly contagious viral disease of cloven-hoofed animals—such as cattle, swine, and

sheep—that causes debilitation and losses in meat and milk production (while FMD does not have human health implications it does have severe economic consequences); and

- plants, such as certain varieties of *Xylella Fastidiosa,* which can kill citrus plants, but does not have human health implications.

Lab levels can also vary depending on their use. For example, research that involves animal or plant pathogens may be designated as animal biosafety levels (ABSL) 1-4 or BSL-3-AG. Similarly, some people may refer to BSL-3 labs as "high-containment" labs and BSL-4 labs as "maximum containment" labs. There are also several types of labs—including clinical, research, teaching, public health (or reference), and production (or commercial)—which are generally categorized on the basis of the work conducted. While these labs all involve work with infectious microorganisms, there are regulatory, accrediting, and risk differences associated with each type. For example, clinical labs within hospitals test patient samples and may often be unaware of the micro-organism they are handling until their tests have identified it. In contrast, research, reference, and production (commercial) labs, while they each have different purposes and environments, tend to be aware of the micro-organisms they are handling. Clinical labs also have specific accrediting and state reporting requirements, and their control structure for handling illnesses is different from other types of labs. We use the general term "biological lab" to include biological labs of all levels or types that handle micro-organisms or clinical samples. We use this general and inclusive term because SRSs could be used in any environment with safety risks, including different types or levels of labs. However, this does not necessarily imply that a single SRS is appropriate or applicable to all labs of varying type or level, although an SRS that encompasses the largest view of a domain as possible has significant advantages. For example, one national SRS would provide information that can cross boundaries where common and similar practices exist and avoid the "stove-piping" of safety information.

Many different federal agencies have some connection with biological labs. Such agencies are involved with these labs in various capacities, including as users, owners, regulators, and funding sources.[12] The CDC and APHIS regulate entities[13] that possess, use, and transfer select agents and toxins.[14] In addition, entities are required to report the theft, loss release of any select agent or toxin to the CDC or APHIS, although we had found reporting failures at some labs subject to this requirement. [15]

Along with environmental, storage, and equipment configurations, various guidelines for lab practices support worker and public safety. These biosafety guidelines offer general and agent-specific containment and risk assessment practices. For example, the BMBL suggests microbial practices, safety equipment, and facility safeguards that vary by type of agent and intended use. These documents are updated periodically—the BMBL is currently in its fifth edition—in order to "refine guidance based on new knowledge and experiences and to address contemporary issues that present new risks that confront laboratory workers and the public health."[16] While the BMBL and other guidelines are useful for promoting safety, they also recognize there are unknown and emerging laboratory safety risks and that ongoing efforts to gather information about those risks is essential for continued safety improvement. One of the key information sources for these updates is published reports of LAIs. However, it is widely recognized that these reports reflect only a fraction of actual LAIs.

To develop evidence-based guidelines and safety-improvement initiatives, other industries with inherent risks to workers and the general public— such as aviation, commercial nuclear power, and health care—collect and analyze safety data. These data can come from safety events. Safety event levels—depicted in terms of a risk pyramid (see fig. 1)—increase in severity as they decrease in likelihood. Whether and where the lines are drawn—between accidents (fatal or nonfatal), incidents, and hazards— varies (1) across industries and (2) according to whether the safety event resulted in no ill effects, minor injuries, or severe injuries or deaths.

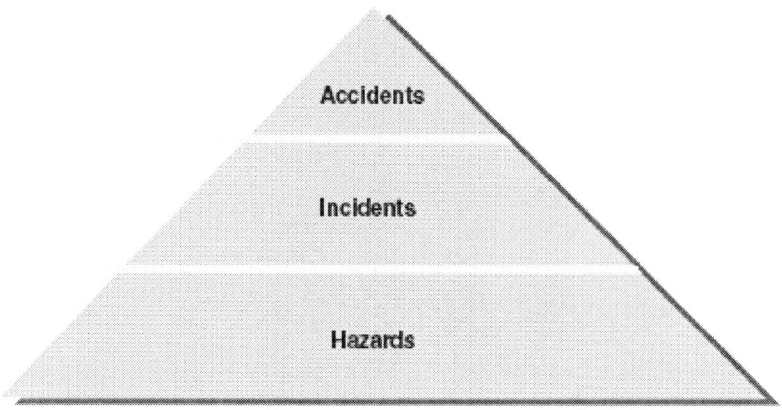

Source: Based on the Heinrich Pyramid.

Figure 1. The Risk Pyramid for Safety Events.

Events at the top of the pyramid—generally identified as "accidents" (sometimes further divided depending on fatality)—have significant potential for harm or result in actual harm to one or more individuals. These events can include radiological exposure, industrial chemical spills or explosions, airline crashes (with or without loss of life), patient medication errors that result in illness or death, and LAIs. Accidents— especially fatal ones—are generally infrequent, hard to conceal, and often required to be reported. Events at the center of the risk pyramid— generally referred to as "incidents"—are those that could have resulted in serious harm but did not. Incidents occur more frequently than accidents and include near misses, close calls, or other potential or actual adverse events and violations, although definitions vary within and across industries. For events at the base of the pyramid—generally referred to as "hazards"—no incident or accident need occur. These events include observations about the work environment, procedures, equipment, or organizational culture that could be improved relative to safety.

Safety data from accidents, incidents, and hazards provide the source information for analysis of accident precursors—the building blocks of events that can lead to injury or death. The focus on precursor data arose as a result of the limited amount of data that could be identified from accident investigations. Such data are often "too sparse, too late and too statistically unreliable to support effective safety management."[17] In addition, the severity and sometimes fatal consequences of accidents often preclude investigators from gathering sufficient detail to fully understand systemic (as opposed to individual) causes of the accident. Incident data are a particularly rich source of precursor information because incidents occur more frequently than accidents. Moreover, incidents do not often rise to the level of regulatory or legal violation because no serious harm has occurred. Workers are therefore generally less fearful of punishment in reporting their mistakes at this level.

Collection of safety data and analysis of accident precursors focus on trying to identify systemic, rather than individual, causes of error. Industries often take this system-based approach to risk management because they recognize that "blaming problems on 'human error' may be accurate, but it does little to prevent recurrences of the problem. If people trip over a step x times per thousand, how big must the x be before we stop blaming people for tripping and start focusing on the step?"[18] The system-based approach focuses on analyzing accident precursors to understand "how and why the defenses failed."[19] According to this approach, blaming individuals for accidents—as in the person-based approach—not only fails to prevent accidents, but also limits workers' willingness to provide information about systemic problems. When precursor information from accidents, incidents, and hazards are analyzed as part of a system, evidence-based, industrywide safety improvements are possible. For example, analysis of reports of health care workers improperly medicating patients has helped identify and address systemic problems with medication labeling and storage. In such cases, hospitals could have punished an individual for the error. Instead, they focused on learning rather than blame, which encouraged worker reporting and led to needed changes in medication labeling and storage. This, in turn, improved patient safety because any health care worker—not just the one that reported the error—will be less likely to improperly medicate patients in the future.

SRSs—both mandatory and voluntary—are the key tool for capturing detailed safety data. Many industries have recognized that the costs of repeated accidents or managing the aftermath of an accident can far outweigh the costs to establish and maintain a reporting system.[20] Despite vast differences across industries, the sources of risk—humans, technology, and environment—are the same. Consequently, the tools— such as SRSs—that industries other than biological labs use to understand these risks can also support evidence-based, industrywide biosafety improvement efforts. This is especially significant in understanding the risks in biological labs because current biosafety guidelines are based on limited information.

While individual states or labs may have reporting mechanisms, no formal system exists for sharing data among all labs. In addition, while data reported through academic journals or state disease registries is accessible industrywide, there are significant reporting barriers. For example, before information about an incident becomes available to others through academic publications, infections must be recognized as laboratory-acquired, deemed scientifically interesting, written up and submitted for peer review, and accepted for inclusion in an academic journal. Furthermore, concerns about losing funding or negative publicity can create barriers to an institution's willingness to encourage publication of LAI information.[21] Reports of infections through state disease registries are also limited because information about the source of the infection is generally not collected and not all infectious diseases are required to be reported. In addition, the infected individual must see a health practitioner who recognizes the status of the disease as reportable and takes steps to report it. Finally, releases without infection—or without recognized infection as a result of a release—are unlikely to be reported at all, despite the valuable precursor data that could be gleaned from the event.

A system for collecting safety data from across the lab community has been proposed as a means to improve the evidence base for biosafety guidelines. However, as indicated by reporting lapses to the mandatory system for theft, loss, and release of select agents, implementation of a reporting system does not immediately create a highly useful one, to which all workers instantaneously submit data on their errors. Finally, when initiating any reporting system, it is important to consider up front and throughout a myriad of design and implementation issues so as to ensure the system is operating as effectively as possible. Consequently, we look to research and experience to inform design and implementation choices.

PROGRAM GOALS AND ORGANIZATIONAL CULTURE GUIDE SAFETY REPORTING SYSTEM DESIGN AND IMPLEMENTATION IN THREE KEY AREAS

Source: GAO analysis of SRS evaluation literature.

Figure 2. Relationship of Program Goals, Organizational Culture, and the Three Key Areas and Subcategories.

According to lessons from our review of the literature,[22] the design and implementation of an effective safety reporting system (SRS) includes consideration of program goals and organizational culture for decisions in three key areas: reporting and analysis, reporter protection and incentives, and feedback mechanisms. Each of the key areas contains subcategories of related decision areas, which should also tie into program goals and organizational culture. Figure 1 illustrates the relationship among program goals, organizational culture, and the three key areas with associated subcategories.

Program Goals and Organizational Culture

A program can have a variety of goals in the design and implementation of an SRS, apart from the primary goal of improving safety, according to the literature. For example, an SRS can be used for regulatory purposes or for organizational learning—a distinction that will fundamentally affect design decisions, such as whether reporting will be mandatory or voluntary, what types of reporter incentives and protections should be included, who will analyze SRS reports, and what feedback will be provided. An SRS can be designed and implemented to meet a variety of subgoals as well. Subgoals can include capabilities for trend analyses, accountability improvement, liability reduction, and performance indicators. The overall goals and subgoals should be determined in advance of design decisions, so that decisions in the three key areas support program goals. Identification and agreement on program goals is best accomplished through the involvement of appropriate stakeholders, such as management, workers, industry groups, accrediting bodies, and relevant federal entities, according to the literature.

Even with well-defined goals, the success of any SRS is intertwined with the organizational culture in which it will operate. Organizational culture—the underlying assumptions, beliefs, values, attitudes, and expectations shared by those in the workplace—affects implementation of programs in general and, in particular, those designed to change that underlying culture.[23] SRSs are fundamentally tools that can be used to facilitate cultural change—to develop or enhance a type of organizational culture known as a culture of safety. A culture of safety implies individual and organizational awareness of and commitment to the importance of safety. It also refers to the personal dedication and accountability of all individuals engaged in any activity that has a bearing on safety in the workplace.[24] Development of a positive safety culture often involves a shift in how workers view and address safety-related events. This shift is supported by data on safety-related events provided by SRSs.[25] Accordingly, an environment in which workers can report safety events without fear of punishment is a basic requirement for a safety culture and an effective SRS. In addition, an important consideration in design and implementation is where on the safety culture continuum an organization is currently positioned and where it would like to be positioned. It is unlikely that workers would report safety events in organizations with punishment-oriented cultures—where workers are distrustful of management and each other. To promote reporting in such environments, systems can be designed with features that help alleviate these worker concerns. However, understanding where the organizational culture is in relation to reporting is essential for choosing system features that will address these concerns.

Changing organizational culture is also generally recognized as a long-term effort that takes at least 5 to 10 years. In high-risk industries, reporting systems are often developed in conjunction with other efforts to make safety a priority, and as the culture changes from these efforts, so might the reporting system to reflect the changing culture. For example, as safety events become more visible or well-defined, reporting forms or requirements can be modified to reflect this new understanding. Similarly, if reporting is waning but safety events continue to occur, adjustments to reporting incentives, definitions of events, and other features may be necessary to improve reporting. Such ongoing assessment of organizational culture can also help identify areas where system adjustments are needed and support efforts to evaluate the contributions of the SRS to safety culture improvement.

Lessons from the literature on the role of program goals and organizational culture in SRSs include the need to

- define overarching program goals and subgoals up front;
- involve stakeholders (e.g., management, industry groups, associations, and workers) in developing program goals and designing the SRS to increase support among key populations;
- assess the organizational culture to guide system design choices in the three key areas; and
- ensure that reporters and system administrators receive adequate training regarding the function and application of the reporting system.

As with any tool for cultural change, the value of the SRS will be commensurate with the investment in its use. If an SRS is to support overall safety improvement, training, outreach, and management support are necessary to instruct staff in the desired culture and use of the new system.

First Key Area: Reporting and Analysis

Among the first design decisions for an SRS are those that cover reporting and analysis. Decisions in this key area include basic questions about the (1) level of event that should be reported to the system, (2) classification of events, (3) report format and mode, (4) management of reporting, and (5) analysis of the reported data.

Level of Event: The Severity of Events Captured Generally Determines Whether an SRS Is Mandatory or Voluntary

The severity of events can vary from safety concerns to mass casualties, and what is considered a "reportable event" has implications for whether reporting should be mandatory or voluntary. Mandatory reporting is generally preferred when program goals are focused on enforcement. Serious events—such as accidents resulting in injuries or deaths—are typically the level of event collected in mandatory SRSs. Mandatory reporting is also generally preferred where there is potential or realized association with injury or death and related regulatory and legal implications, as in accidents. Voluntary reporting is generally preferred when the program goal is learning—identifying actions, processes, or environmental factors that lead to accidents. Voluntary reporting in these cases is more appropriate because the goal is improvement rather than compliance. Events at the incident level—errors without harm, near misses, close calls, and concerns—are less serious than accidents and are typically collected through voluntary SRSs. Both mandatory and voluntary reporting systems are often employed concurrently—sometimes independently and sometimes in complementary roles—because programs face the dual requirements of regulating and promoting safety improvement.

The level of event to be reported also depends on the organizational culture. Industries new to safety reporting—in particular, those in which the definition or recognition of an accident is unclear—may find it particularly difficult to identify a reportable incident or hazard. If the reporting threshold is set too high, significant safety hazards may go undetected and unreported. In such environments, a low initial threshold for reporting might be helpful, raising it over time as workers develop familiarity with reportable events. However, because of the greater frequency of incidents and safety concerns, voluntary SRSs can be overwhelmed by the volume of submitted reports. SRSs that focus on a particular type of incident or hazard area may help to counteract this problem. In addition, if the reporting threshold is set too low, reporters may feel events are too trivial for reporting and that the SRS has little value. For example, surveys of nurses and doctors have shown a range of opinions that constitute a barrier to reporting, including beliefs that not all near-miss errors should be reported or that reporting close calls could result in significant change. The prevalence of these beliefs may reflect that a "reporting culture"—one in which staff recognize and submit reportable events—is not fully established.

> Lessons from the literature on determining the level of event for reporting include the need to
>
> - base the decision for mandatory or voluntary reporting on (1) the level of event of interest and (2) whether the SRS will be used primarily for enforcement or learning and
> - set reporting thresholds that are not so high that reporting is curtailed, but not so low that the system is overwhelmed by the number and variety of reportable events.

Classification of Error: Error Classification Can Guide Reporting and Facilitate Information Sharing, but Can Limit Information Flow if Too Restrictive

To facilitate data-sharing across the organization or industry, classification schemes provide standardized descriptions of accidents, incidents, and concerns. Effective classification schemes can facilitate safety improvement across organizations and industry by providing a common language for understanding safety events and precursors. For example, if several hospitals use a standard classification scheme to submit incident reports to a patient SRS, the resulting data can be used to examine incident data across hospitals. Such data allow benchmarking of similar occurrences and promote a better understanding of core hazards that exist across an industry. Clearly defined and familiar classification terminology can also help workers understand when and what to report. However, achieving a well-defined and clear classification scheme—especially one that can be used across an industry—can be difficult because different groups within an organization or across an industry may classify events differently. For example, one study on medical error reporting found that nurses classify late administration of medication as a medical error, whereas pharmacists do not.

Classification schemes should be broad enough to capture all events of interest, but also well-defined enough to minimize receipt of extraneous information. For example, organizational learning systems, like FAA's NASA-run Aviation Safety Reporting System (ASRS), include a broad definition of safety-related events to facilitate voluntary reporting of all events. Alternatively, mandatory systems may include a more specific classification scheme to capture deviations from standard operating procedures. However, overly restrictive schemes may lead workers to focus on certain events and neglect to report others.

For example, if a classification scheme is developed to consider only compliance with an industry's standard operating procedures, workers may not report safety-related incidents that involve factors other than compliance. Similarly, overly detailed classification schemes may be confusing for reporters if they do not know the appropriate codes to apply. In addition, a classification scheme must be clear enough for workers to understand what counts as a reportable incident. Otherwise, underreporting or misreporting of incidents may result. If possible, use of pre-existing industry-specific terminology in the classification scheme can support information flow across the industry and help workers—especially in industries new to safety reporting—adapt to the SRS. Lastly, a classification scheme may require the flexibility to allow different sites to adapt fields and elements to match their own program goals and organizational cultures.

Design of a classification scheme may incorporate several strategies, including (1) using an existing classification scheme from another SRS, (2) modifying an existing classification scheme for use in a new SRS, (3) developing a classification scheme based on incident reports from the new or a similar SRS, or (4) using experts to develop a classification scheme.

Lessons from the literature on designing classification schemes and associated terms include the need to

- develop classification schemes and associated terms that are clear, easy to understand, and easy to use by drawing on terms already well understood in the industry;
- test whether classification terms are clearly understood by different groups in the organization;

- allow sufficient flexibility to (1) avoid narrowing the scope of reporting in a way that limits all events of interest at the chosen level of event, (2) allow different sites—if multiple sites will be reporting to the same system—to adapt fields and elements to match their own organizational culture, and (3) capture different types of events and precursors, as they can change over time; and
- develop a classification scheme that best suits the analytical requirements and the comfort level of the organizational culture with safety reporting and safety event terms.

Format and Mode: Report Mode and Format Must Balance Needs for Quality and Quantity of Reported Information with Reporter Burden and Proclivity to Report

Reporting must be readily accessible and allow for sufficient description of safety events without overburdening reporters with extensive narrative requirements. Data collection considerations include the format of the report (that is, the types of questions included on the reporting form) and the mode of the report (that is, how a report is physically submitted to the SRS, for example, by paper or Internet). Both the report format and mode can affect the incentive to report; the ease of reporting; and the type, quantity, and quality of data collected. Decisions regarding the format and mode of reporting are closely tied to the type of data desired from the SRS and the organizational culture.

Report formats affect the quantity and quality of reports. For example, question formats that allow workers to explain the incident through narrative description may yield extensive details about the incident. The literacy skills of the reporting population are important considerations as well. Long narratives might be simple for the highly educated but intimidating to those with less writing proficiency. However, if workers are resistant to reporting, structured question formats that use check-boxes or drop-down boxes with categories may decrease the time it takes to complete an incident report and thereby increase the incentive to report. Using structured question formats will also decrease the amount of coding and qualitative analysis that must be performed to examine the data. One limitation of structured question formats, however, is that in industries new to safety reporting, classification terms may not be well developed or understood by the reporting population.

Options for SRS modes include paper, telephone, or electronic or Web-based form. Although Web-based forms may increase the ease with which data are collected, workers may be fearful of entering incident reports using a Web-based form because reports can be traced back to them. If workers perceive that the culture is punitive, mail reports—especially to an outside entity that manages the system—can be the most effective mode choice to alleviate these concerns. However, accessibility of reporting forms can also affect the likelihood of reporting. For example, if paper forms are outside the immediate work area and require effort beyond the normal routine to complete, then reporting may be curtailed. Since many workers have ready access to the Web, a combination of Web and mail reporting may address both access and sensitivity concerns.

Lessons from the literature on format and mode choice include the need to

- base decisions about report formats on (1) the type of data needed for analysis, (2) capabilities of the reporting population, and (3) maturity of existing safety event classification schemes within the industry and
- base decisions about report mode on (1) the accessibility of the mode to the reporting population and (2) workers' concerns about and willingness to report.

Reporting Management: SRS Administration and the Designated Reporting Population Can Affect Willingness to Report and Analytical Possibilities

Reporting management includes decisions about SRS administration—who will collect, analyze, and disseminate reports—as well as decisions about who is allowed to submit reports. The choice of the entity responsible for collecting, maintaining, analyzing, and disseminating may affect the willingness of workers to submit reports. For example, if workers perceive a punitive organizational culture or a lack of confidentiality, they may be unwilling to submit reports to an SRS within the workplace. An SRS managed by an independent, external entity might alleviate these concerns. However, an organization may have better awareness than an outside entity of internal safety issues, expertise in analyzing and addressing them, and mechanisms for encouraging participation in safety reporting. Consequently, decision makers must weigh a variety of culture-related and resource considerations in deciding how to administer an SRS.

The openness of reporting—whether reporting is available to all workers or only to those in select occupations or positions—will also affect the type and volume of data collected. For example, many individuals— including pilots, ground crew, and controllers—can submit reports to FAA's NASA-run ASRS, whereas only airlines can submit reports to the Voluntary Disclosure Reporting Program (VDRP). An open SRS, which accepts reports from different staff levels or occupations, offers the potential for analysis of events from several perspectives. However, such an SRS may be subject to staff hierarchies that can limit reporting among certain employee groups or professions. For example, in the medical industry, even when reporting is open to both doctors and nurses, several studies have shown that nurses have a greater awareness of and are more likely to submit reports to an SRS than doctors. Similarly, reporting may be attenuated if events must be reported up a chain of command, rather than directly by those involved in an event. Direct reporting—regardless of position or occupation—can increase the likelihood of reporting on a particular event.

> Lessons from the literature on system administration and the reporting population include the need to
>
> - base the decision for internal or external system administration on (1) workers' degree of concern over punishment and confidentiality and (2) availability of internal expertise and resources to analyze and encourage reporting and
> - base decisions about who will be allowed to report on (1) awareness of reporting hierarchies and (2) the type of information desired for analysis

Analytical Process: Report Prioritization, Data-Mining Techniques, and Technical Expertise Can Enhance Results

Analytical processes that focus on identifying safety improvements—using report prioritization, data-mining techniques, and safety and industry experts—can enhance the usefulness of reported information. Frequently, the first step in analyzing reported data is determining whether immediate action should be taken to address a safety concern. Subsequently, analyses that explore why a particular event may have occurred—such as root cause analysis—may be used to understand the contributing factors to safety events and to design solutions to the problem. Data-mining techniques, including those that combine safety reports with other databases, can also be used to look for patterns of events across organizations or a broad range of reports. Data mining requires the capability to search for clusters of similar events and reports that share common characteristics. Technical expertise, as well as specialized software, access to other data sources, and data format requirements, affects data-mining capabilities. For example, data-mining searches may be more complicated when error reports include both structured and open text (narrative) formats because open text must be made suitable for data mining. In addition to these retrospective analytical techniques, probabilistic risk assessment methods may also be used as a proactive approach to examine all factors that might contribute to an event. Literature on SRS use in industries, such as nuclear power and aviation, advocate using a combination of these approaches to provide a more thorough analysis of reported data. Finally, using data analysis techniques to prioritize incident reports can facilitate analysis by identifying which reports require further analysis or demand immediate review because they represent serious safety concerns. Because analysts must have the technical skills and relevant knowledge to make sense of the data, decisions about the analysis will be linked with system administration and whether technical and industry expertise reside within the organization. Thorough analysis may require multidisciplinary committees that contribute a variety of expert perspectives, but the breadth of expertise required may not be readily available within an organization. For example, analysis of medication error reports may be conducted through multidisciplinary committees that include physicians, nurses, pharmacists, quality managers, and administrators. In the airline industry, an event review team (ERT), consisting of representatives from the air carrier, the employee labor association, and the FAA, is used to analyze reports as part of the Aviation Safety Action Program (ASAP).

Lessons from the literature on analytical process include the need to

- use a report prioritization process to quickly and efficiently address key safety issues as they arise and
- align analysis decisions with (1) report formats, (2) system administration and location of technical expertise, and (3) availability of other relevant data needed for analysis.

Second Key Area: Reporter Protections and Incentives

SRSs—whether mandatory and voluntary—depend on the willingness of workers to report mistakes they or others have made. It is unlikely that workers would take the risk of reporting without protections that provide confidence that their reports will be kept private and incentives to report their errors. There are a variety of ways to design SRSs to protect the identity of the reporter and to encourage reporting, including (1) accepting anonymous reports, (2) providing effective confidentiality protections on reported data, and (3) deidentifying data sets. The principle reporting incentive is limited immunity—whereby workers are granted protection from certain administrative penalties when they report errors. There are advantages and disadvantages to anonymous and confidential reporting, and decisions about which to use should be guided by program goals and culture-related considerations.

Anonymity Is the Surest Method for Protecting Reporter Identity, but Can Limit Reporting Data

Anonymity—reporting without identifying information—protects reporters against legal discovery should the data be requested in a subpoena. Because an individual's name is not tied to an incident report, anonymity may lower the psychological barrier to reporting, including fears about admitting a mistake or looking incompetent, disclosure, and litigation. Anonymity may be critical in motivating reporting among workers in an organizational culture seen as punitive, especially when legal protections for reporter confidentiality may not be feasible or well established. Report mode is also linked with reporter protection choices. For example, one SRS for medication errors was developed as a paper-based system because administrators felt any electronic system could not be truly anonymous.

Despite the protection anonymity offers reporters, there are distinct disadvantages, including the inability to obtain clarification or further information from reporters. This limitation may compromise the integrity of system data because investigators have no means for validating and verifying the reported information. In addition, anonymous data sets tend to be less detailed than identified data sets. Initial reports from identified data sets can be supplemented by follow-up interviews with reporters. The need to follow up with reporters may also make anonymous reporting unfeasible, even in organizations where significant reporting concerns exist. Anonymous reporting also tends to limit the number of data elements that can be derived from reports, making these data sets less useful than others, particularly when trying to identify patterns of error. For example, if fields that could identify reporters—such as occupation, location, and position—are not collected, statistics on safety events across organizational subunits or occupations would be impossible.

Another disadvantage of anonymity is that reporters cannot be contacted for clarification or to provide direct feedback—a useful technique for obtaining worker buy-in to the system. If reporters are given specific feedback on actions taken to address issues brought up in their reports and the outcomes of these actions, then reporters are more likely to (1) attribute value to the SRS and (2) continue submitting reports. Some SRSs have addressed this problem by offering a compromise. Reporters can receive a unique identification number that allows them to track the progress of their reports through the SRS. However, if reporters are mistrustful enough that anonymous reporting is necessary, they may not feel comfortable using an optional identification number provided by the SRS. Even anonymity may not be enough to alleviate reporters' fear of retribution. Other disadvantages of anonymous reporting include the potential for (1) workers to falsely report on the behavior of others in the absence of report validation and (2) managers to discredit information about concerns or incidents as reports of "troublemakers." Yet another disadvantage is the inability to maintain anonymity in small reporting populations or where the circumstances surrounding an incident are so specific (to an organization, individual, date, and time) that any mention of them would disclose the parties involved.

Confidentiality Enables Follow-up with Reporters but Includes the Potential for Compromising Reporter Identity

Confidential reports allow investigators to follow up with reporters to gain a better understanding of reported incidents because the link between the reporter and report is maintained. However, fear of providing identifying information may limit reporting. Confidentiality is accomplished through legislative, regulatory, or organizational provisions to protect reporter privacy. Such provisions can include exemptions from subpoena or disclosure, protections against civil or criminal lawsuits for reporting, or criminalizing confidentiality breaches. For example, some state-based mandatory SRSs for medical errors include statutory provisions that protect reporters from some potential legal liability. One international aviation SRS has legislation making confidentiality breaches a punishable offense.

Maintaining identifying information enables data analysis across professions and organizations, which can aid in benchmarking. Such information can reveal whether recurring incidents indicate problems within a specific organization or profession as opposed to those that are industrywide, thereby targeting interventions to areas in greatest need. Reporting formats may be less burdensome for confidential systems than for anonymous systems, which must gather all details up front. Confidential reporting allows investigators to gather significant information through follow-up interviews, so less detail needs to be provided on the reporting form. In the literature, report follow-up was associated with a variety of positive results. For example, it can (1) add to reporters' long-term recall of the event, enhancing the quantity and richness of information collected; (2) support event validation and clarification; and (3) bring closure to an incident and assure reporters their information is being taken seriously, thus increasing the likelihood of future reporting.

A potential disadvantage of a confidential SRS is that workers may be fearful of the consequences—real or implied—of reporting. Moreover, for systems untried by the legal system, the surety of confidentiality provisions can be—in reality or perception—tenuous. For example, the Applied Strategies for Improving Patient Safety (ASIPS) is a multi-institutional reporting system designed to analyze data on medical errors and is funded by the Agency for Healthcare Research and Quality (AHRQ). This voluntary SRS for patient safety events relies on confidential reports provided by clinicians and office staff. While this reporting system promises reporters confidentiality within the system, the program can offer no protection against potential legal discovery. However, because ASIPS is funded by AHRQ, ASIPS reporters would be protected by the confidentiality provision in AHRQ's authorizing legislation, although the protections provided by this provision have never been tested through litigation. Because of the uncertainty of confidentiality protections, administrators of ASIPS chose to build strong deidentification procedures—removal of identifying information from reported data—into the system rather than rely solely on confidentiality protections. Another potential disadvantage of confidential SRSs is that costs may be higher than an anonymous system if follow-up interviews with reporters are part of SRS requirements. Sufficient resources are required for investigation and follow-up with reporters; however, resource constraints may limit these actions. Additional resource commitments (in the form of follow-up interviews) are also assumed by those who submit confidential reports.

Data Deidentification Provides Additional Reporter Protection

Data deidentification supports confidentiality provisions since the deidentification process makes it difficult to link reports to specific individuals or organizations. Deidentification can also support feedback mechanisms because the data can be readily shared within and across organizations and industries. Data can be deidentified at the source or in summary reports and data systems. Source deidentification involves removal and destruction of all identifying information from reports after follow-up and investigation have been completed. Secondary data deidentification involves removal of identifying information in summary reports or databases for sharing safety information and alerts. Deidentification of source reports strengthens confidentiality protection because records are unavailable even if they are subpoenaed. Source report deidentification may require (1) technical solutions if reports are collected electronically and (2) special processes if collected in another format. Eliminating the link between the reporter and the report can help reinforce the confidential nature of an SRS and provide an incentive for reporting, as long as the process for deidentification is understood by the reporting population. Deidentified data can be readily shared within or across organizations and industries, enhancing analytical possibilities by increasing the number of reported incidents available for analysis.

Limited Immunity Provides Reporting Incentive

Limited immunity provisions can increase the volume of reports, particularly when there are emotional barriers, such as fear about reporting one's mistakes. These provisions offer protection from certain legal or regulatory action if certain requirements are met. For example, the ASRS offers limited immunity from enforcement actions provided certain requirements are met and the incidents do not involve criminal or negligent behavior. The literature suggests that the immunity provisions offer a strong incentive to report and that pilots would not submit ASRS reports if these provisions did not exist. Numerous international SRSs also contain immunity provisions, including the Danish aviation SRS and patient care SRSs in both Australia and Israel.

Lessons from the literature on choosing reporter protections and incentives include the need to

- base the choice between anonymity and confidentiality on (1) organizational culture, especially workers' degree of concern about punishment and confidentiality, and (2) the amount of detail required for analysis and whether it can be collected without follow-up;
- consider hybrid systems in which confidential and anonymous reporting are used simultaneously if there is a conflict between organizational culture and data need;

- develop data deidentification measures to support confidentiality and data-sharing efforts; and
- consider limited immunity provisions to increase the reporting incentive.

Third Key Area: Feedback Mechanisms

Because a primary SRS function is safety improvement, the system must include feedback mechanisms for (1) providing actionable safety information to the relevant populations and (2) improving the SRS through identification of reporting gaps across occupations or locations and evaluation of the effectiveness of the system as a safety tool.

Feedback to Reporters and Industry Promotes Safety Improvement and Reinforces Reporting

To support its primary function of safety improvement, an SRS must include feedback mechanisms for providing actionable safety information to the relevant populations.

A variety of populations can benefit from SRS feedback, including (1) reporters, (2) managers, (3) organizations and the industry at large, and (4) system administrators.

Feedback to reporters is essential in order to promote safety and reinforce the benefits of reporting.

If workers who report safety events do not see any evidence that their report has been used, they may question the value of the system and discontinue reporting.

Feedback among managers promotes management awareness of safety concerns, management buy-in, and top-level efforts to address those concerns. Feedback across the organization or industry can provide tangible evidence of the value of the SRS by alerting management and workers to important safety issues.

Industry feedback can also provide a benchmark to compare safety across similar organizations when data are (1) collected at the local level and (2) compiled in a centralized regional or national database. Use of such benchmarks may help decision makers identify gaps in performance and practices that may improve safety conditions in their own organization.

Feedback on System Performance Supports Targeted Outreach and System Improvement

Feedback mechanisms for system evaluation are also important in ensuring the SRS's continued effectiveness. Feedback on reporting gaps across occupations or locations can help identify nonreporting populations. When these reporting gaps are compared with other data— such as reports from comparable sites—they can help identify areas in need of targeted outreach and training. In addition, feedback from safety culture and system-user surveys, which assess safety and reporting attitudes, can be used to evaluate the effectiveness of an SRS. Performance metrics on safety improvement can be incorporated into these surveys, providing information on the degree to which program goals are being met and identifying areas of needed system improvement.

Lessons from the literature on choosing feedback mechanisms include the need to

- provide direct feedback to reporters to foster worker-specific buy-in for reporting;
- provide regular, timely, and routine feedback—for example, in the form of newsletters, alerts, Web sites, and searchable databases—to support overall organizational buy-in for reporting;
- provide positive feedback to managers who receive a high volume of reports to demonstrate the importance of reporting and counteract the perception that error reporting reflects poorly on management;
- use the data to identify reporting gaps for targeted outreach and training; and
- evaluate the effectiveness of the SRS to support ongoing modification and improvement.

Case Studies Demonstrate the Need for Assessment and Resources in Design and Implementation and Suggest Certain Features in the Three Key Areas

Lessons from case studies of safety reporting systems (SRS) in three industries—aviation, commercial nuclear power, and health care— indicate the importance of cultural assessment and resource dedication in SRS design and implementation, and suggest certain features in the three key areas.[26] Although the industries differ in type of work, regulation, and ownership, all three face substantial inherent risks to health and public safety and have made significant investments in promoting safety through voluntary SRS programs. Consequently, their experiences suggest lessons that can be applied to the design and implementation of an SRS for biological labs. Collectively, these SRSs reflect 70 years of safety reporting experience.

In particular, the FAA's NASA-run Aviation Safety Reporting System (ASRS) in aviation, the Institute of Nuclear Power Operation's (INPO®) Significant Event Evaluation—Information Network (SEE-IN®) system in commercial nuclear power, and VA's internally managed Patient Safety Information System (PSIS) and NASA-run Patient Safety Reporting System (PSRS) in VA health care provide the basis for the following four lessons for SRS design and implementation:[27]

1. Assessment, dedicated resources, and management focus are needed to understand and improve safety culture.
2. Broad reporting thresholds, experience-driven classification schemes, and processing at the local level can be useful SRS features in industries new to safety reporting.
3. Strong legal protections and incentives encourage reporting and help prevent confidentiality breaches.
4. A central industry-level entity facilitates lesson sharing and evaluation.

Lesson 1: Assessment, Dedicated Resources, and Management Focus Are Needed to Understand and Improve Safety Culture

The case studies demonstrate that establishing a robust safety culture is neither quick nor effective without a multipronged effort—involving assessment, dedicated resources, and management focus—to recognize safety challenges and improve safety culture. Despite the costs and challenges of implementing an SRS, the industries recognized they could not continue to operate without safety improvements and their SRSs were a key tool in these efforts.

Assessing Safety Culture Can Alert Management to Workplace Safety Issues

Each of the three industries created its SRS after recognizing that existing operations and safety culture posed an unacceptable risk to workers and the public. In both the aviation and the commercial nuclear power industries, SRS initiation was prompted by serious accidents rather than a proactive assessment of the safety culture. The Veterans Health Administration proactively initiated an SRS program after its administrators and patient safety advocates recognized the need to redesign systems "to make error difficult to commit."[28] Such assessments can reveal systemic safety culture problems before they become critical.

Aviation

The concept of a voluntary aviation reporting system was suggested in 1975 by the National Transportation Safety Board (NTSB), the FAA, and the aviation industry following an investigation of a fatal airline accident near Berryville, Virginia. The NTSB found that the accident might have been averted if previous crews' reports about their near-miss problems in that area had been shared. These problems included inadequate aviation maps and the cockpit crews' misunderstanding related to the air traffic controllers' terminology. The NTSB reported that the industry culture made it difficult to report these problems. These cultural barriers were apparently known, although a safety culture assessment might have afforded proactive efforts to correct them. As one solution to these problems, the NTSB suggested an aviation SRS, initially managed by the FAA and known as the Aviation Safety Reporting Program. But within a few months, the FAA had received few reports. It therefore transferred operation and management of the program to NASA and renamed it the Aviation Safety Reporting System (ASRS).[29]

Commercial Nuclear Power

In 1979, the partial meltdown of a reactor at Three Mile Island (TMI) in Pennsylvania led to the creation of INPO, an industry-initiated technical organization that collects, studies, and shares safety lessons throughout the industry using the SEE-IN program. The INPO program was developed and is managed independently of the Nuclear Regulatory Commission (NRC) regulatory requirements. Although the NRC regulates the safety of commercial nuclear power generation,[30] at the time of TMI, nuclear utilities had been operating with a high degree of autonomy and were fairly insular, according to a 1994 study.[31] The 1994 study of the safety culture at nuclear reactors found that the management style reflected the culture of conventional energy plants—a “hands-off management” and “fossil fuel mentality” that emphasized maximum energy production as the highest value.[32] An industry official explained that the TMI accident was a shock for the industry, which became determined to operate its nuclear reactor facilities safely and reliably, thereby convincing the American public it could be responsible and safe. The entire U.S. commercial nuclear power industry joined INPO within months of the TMI incident, and remains members today. The industry focused early efforts on plant evaluations to understand the culture that had led to the TMI accident. Within a year, INPO produced the first of its Significant Operating Event Reports, which provide information on identified safety problems and make recommendations for improvement.

Despite safety advances in the decades after INPO was established, the industry was once again reminded of the importance of safety culture assessment in 2002, when corrosion ate a pineapple-sized hole in the reactor vessel head at the Davis-Besse plant in Ohio.[33] Prior to this incident, INPO had given individual plants the responsibility for assessing their safety culture—assuming that they had a good understanding of it. Investigation revealed that a weak safety culture contributed to the incident. After the Davis-Besse incident, INPO re-emphasized the importance of proactively assessing safety culture before critical safety failures occur. In response to the incident, they recommended that safety culture assessments be a permanent, periodic requirement.

Health Care

After VA hospital accidents that had resulted in harm to patients, the VA established the National Center for Patient Safety (NCPS) in 1999. That unit designed and launched two options for reporting—one internal (the PSIS) and one contracted (the PSRS) to the same NASA center that operates ASRS for the FAA.[34] The VA launched its SRS program guided by a vision emerging in the medical community to “create a culture in which the existence of risk is acknowledged and injury prevention is recognized as everyone’s responsibility.”[35] The VA hired management with experience in NASA’s safety programs, who surveyed safety culture as they initiated the SRS. In addition, the NCPS has conducted three nationwide safety culture surveys, beginning in 2000, to understand the attitudes and motivations of its frontline workers.

The most recent, in 2009, allowed the NCPS to identify a subcategory of caregivers for intervention.

Improving Safety Culture Requires Dedicated Resources, Including Time, Training, and Staff Investment

Safety culture improvement depends on a robust reporting culture, which requires considerable investment of time and resources. As the experiences of the three industries demonstrate and as shown by SRS data from two of the case industries, these investments pay off in an increase, over time, in the volume of safety reports.

Figure 3 illustrates time frames and growth in SRS reporting for FAA's ASRS and the VA's PSIS.

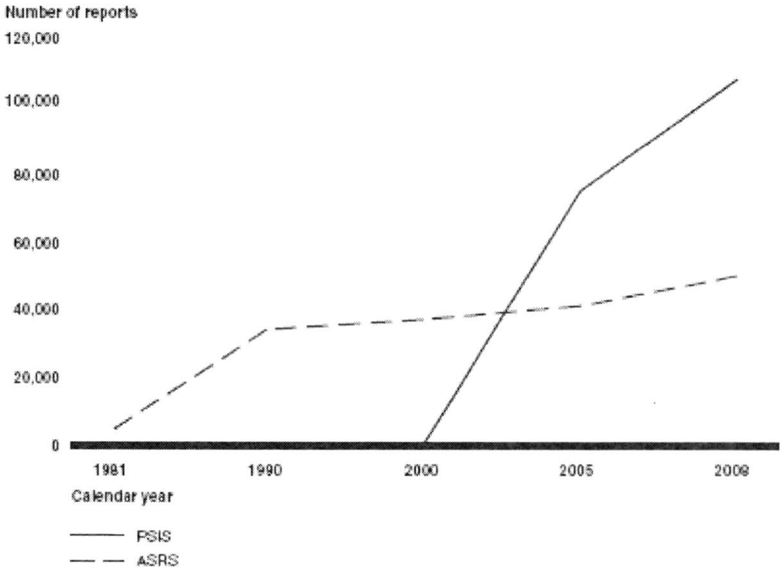

Source: VA, NASA.

Note: Comparable data from the commercial nuclear power industry are not available. The earliest data for the ASRS are in 1981, although the system began in 1976.

Figure 3. Growth in Aviation and VA Health Care Safety Reporting, 1981 to 2008.

Through conventional classroom and seminar training, workers in some industries learned the terms, goals, and instruments of the new voluntary SRS. Several innovative training opportunities were also marshaled, including on-the-job training and employee loan and training programs focused on improving teamwork. Both types of training supported safety culture change and developed trust in the SRS. Staff time and investment at all levels were necessary to accomplish these training goals.

Aviation

From the inception of ASRS, the volume of aviation safety reports grew slowly, indicating an increasing understanding among reporters of the multiple factors that contribute to safety. However, a 1994 National Academy of Public Administration (NAPA) evaluation, requested by the FAA, found that FAA funding provided to NASA for the operation and management of the ASRS had not kept pace with the work.[36] According to a NASA ASRS official, because resources were insufficient to perform a detailed analysis on all the reports, reports are triaged. Only those deemed most hazardous receive deeper analysis. The NAPA report also noted that the aviation community broadly affirms the safety value of ASRS and uses the data for training and safety awareness. By contrast, some FAA line employees said ASRS was of limited use. As a result of the NAPA report and congressional actions, the FAA modestly increased funding. After the NAPA recommendation to modernize, the ASRS transitioned from paper to electronic report submissions. A recent FAA-sponsored study recognizes the importance of training and retraining all SRS stakeholders, offering best practices for formal and informal training. Reporting has increased. ASRS currently receives about 50,000 reports per year, which demonstrates a sustained level of trust in reporting. However, the study of best practices in FAA's voluntary reporting options recommended that SRS managers assess the availability of resources and plan for acquiring them, as resource needs are likely to increase over time.[37] In further recognition of the importance of resources to ASRS, the latest Memorandum of Understanding between the FAA and NASA also includes a yearly inflation factor for the ASRS budget.

Commercial Nuclear Power

Safety reporting to INPO's SEE-IN program began in 1980. The volume of reports forwarded to INPO from the plants is between 3,000 and 4,000 annually.[38] Early safety reports tended to focus on technical failures and INPO realized that reporting on human error needed to increase, according to an INPO liaison.39 Moving beyond reporting equipment failure required significant training. To encourage reporting of both equipment and human factor issues, INPO established and continues to accredit training courses. Recognizing the importance of having staff with industry knowledge to communicate the relevance of safety and reporting in a way that is palatable to industry, INPO began a second wave of hiring of people with nuclear industry experience to ensure the safety science message was managed and communicated in a way that both sides could understand. Despite increases in reporting, however, the Davis-Besse incident in 2002 highlighted the serious consequences of lapses in safety culture. Among other actions, INPO issued its safety principles document in 2004, which provides a framework for assessing safety culture. The document outlines aspects of positive safety culture, such as workers' questioning attitudes that support reporting and managers' demonstrated commitment to safety through coaching, mentoring, and personal involvement in high-quality training.

Health Care

Reporting to the VA's PSIS grew strongly, from 300 incidents reported annually at local hospitals in 2000 to 75,000 in 2005. Yet, the initiation of a voluntary safety reporting system in the VA health care facilities has faced considerable cultural and institutional challenges. For example, one study found the various professions within hospitals disagreed—when presented with scenarios such as late administration of medication—as to whether an error had occurred. In congressional testimony in 2000,[40] we had observed that if the VA hospital system was to implement an SRS, the VA would face a challenge in creating an atmosphere that supports reporting because hospital staff have traditionally been held responsible for adverse patient outcomes. In our 2004 report, we also found that power relationships, such as nurses reluctant to challenge doctors, can be obstacles to patient safety. However, after the first 3 years of the VA health care system's SRS, the cultural change that supports safety reporting was under way at three of four facilities studied, as a result of experiential training in addition to conventional classroom training. The growth in reported events to the VA SRS over the last 10 years and our 2004 study suggest that the actions that the VA took can be successful in supporting a safety culture and reporting.

Experiential—that is, on-the-job—training, in addition to conventional classroom experience, fostered the habit of reporting safety events at many VA hospitals. Since the initial years of the VA's hospital SRS, clinicians and other VA workers have been selected to participate in the hospital-based analysis of SRS reports so that they could learn how the reports would be used. Once patient safety managers prioritized reports, interdisciplinary teams of hospital staff, including local frontline clinicians, looked for underlying causes and devised systemic fixes. Through this experience, clinicians and other hospital staff saw first-hand the rule-driven and dispassionate search for root causes that resulted in a systemic fix or policy change rather than punishment. We found that (1) this training fostered a cultural shift toward reporting systemic problems by reducing fear of blame, and (2) staff were impressed with the team analysis experience because it demonstrated the switch from blame and the value of reporting close calls.[41] In addition, the VA brought together facility-level workers, including patient safety managers from VA medical centers across the nation, to introduce them to the SRS. Through these seminars, staff were introduced to SRS terms, tools, goals, and potential obstacles. They heard success stories from industry and government, findings from the early VA safety culture surveys, and recent alerts and advisories.

Changing Safety Culture Requires Management Focus

To overcome cultural barriers to safety reporting—such as fear of punishment, lack of trust between coworkers and management, and hierarchical prohibitions on communication—management demonstrations of support for the SRS are important. In the three industries, this support was demonstrated through the deliberate use of tactics shown to be effective at changing safety culture and supporting safety reporting such as (1) open communication across the workplace hierarchy encouraged in small group discussions and meetings with managers; (2) storytelling, a tool to direct changes in norms and values; and (3) rewards for participation in safety reporting or open communication in meetings.

Aviation

The three decades of ASRS experience demonstrate the importance of consistent focus versus episodic efforts to publicize and support the SRS. In the early stages of ASRS implementation, the FAA and ASRS staff relied on small group briefings and promotional documents to foster awareness and trust in reporting. For example, the FAA, through its Advisory Circular, notified the aviation community that the system was operational and, along with NASA, issued press releases and conducted briefings about the system. In addition, industry groups and airlines publicly expressed support for the system, and, according to a 1986 NASA report, an advisory group carried "the word about ASRS program plans and accomplishments back to their respective constituencies."[42] Other early promotional efforts included the distribution of descriptive brochures and posters to operators, FAA field offices, air traffic control facilities, and airline crew facilities. As a result of these efforts, according to NASA's 1986 report, the number of reports coming into the system in the early years exceeded expectations. However, a NAPA study 8 years later raised concerns about the lack of publicity. That study found that pilots lacked knowledge of the ASRS and the immunity features[43] and questioned the FAA's credibility. NASA responded with a second promotional surge by (1) publishing its first *CALLBACK,* a monthly online bulletin, and (2) touring FAA regional headquarters to promote the SRS. However, the NAPA study concluded that the lack of internal FAA support for the ASRS had limited the degree to which FAA uses ASRS data, and led to questioning the legitimacy of ASRS products and activities. That study also found that FAA line officers (with the exception of the Office of Aviation Safety) thought the ASRS had limited utility, and some even suspected bias in reporting as a result of reporters' interest in earning immunity from FAA enforcement actions. To address these concerns, the FAA has recently been advised to elevate the importance of establishing an initial shared vision among all stakeholders through open discussion and training and sustained promotion efforts.[44]

Commercial Nuclear Power

INPO focused on leaders and employee loan programs to change the industry's safety culture one employee and one plant at a time. Leadership's demonstrated commitment to safety is a key INPO principle for a robust safety culture.

This key principle stems from the philosophy of having "eyes on the problem." That is, plant managers must be out in the work areas, seeing things and talking to employees in order to reinforce a safety culture. This principle also includes reinforcing standards and encouraging candid dialogue when safety issues arise. Such reinforcement can be in the form of rewards for reporting, such as being congratulated at plant meetings for a "good catch." Managers also have incentives to encourage workers to report. Following its biannual inspections, INPO summarizes its assessment of the plant's safety conditions, providing a numeric score, partly based on the robustness of the plant's SRS.

These safety scores are important to plant managers because they can affect regulatory oversight and insurance premiums. Scores range from 1 to 5, with 1 as the top safety rating. While these assessments may result in more attention and assistance for safety improvements, they also instill pride in the plant, and at annual managers' meetings, managers of plants with the highest scores receive recognition.

INPO has also facilitated active peer review and employee loan programs to break down the insularity of the TMI era. When individuals with in-depth industry experience participate in the inspection process and work at INPO headquarters, they see firsthand the excellence other plants practice and how those practices relate to INPO safety initiatives.

Health Care

The VA hospitals used small group meetings, storytelling, and small rewards to reinforce safety reporting. At the most successful VA hospital we reviewed in 2004, administrators held more than 100 small group meetings where storytelling was used in order to introduce the new SRS.[45] VA hospital administrators used examples from aviation wherein two airline pilots failed to communicate well enough to avoid a fatal crash. The crash might have been avoided had the first officer challenged the captain. This story raised parallels with the medical hierarchy and led to discussions about similar unequal power relationships in the hospital. Administrators introduced more effective ways to challenge authority, naming it "cross-checking." An early report to the VA SRS, which involved nearly identical packaging for an analgesic and a potentially dangerous drug, was made into a poster as part of the campaign for the SRS. The more successful VA hospitals rewarded the month's best safety report with a plate of cookies or certificates to the cafeteria. This playful openness reduced secrecy and fears of punishment and increased comfort with reporting, according to our 2004 analysis.

Lesson 2: Broad Reporting Thresholds, Experience-Driven Classification Schemes, and Processing at the Local Level Are Useful Features in Industries New to Safety Reporting

After the three industries instituted a voluntary SRS, workers experienced a sharp learning curve in recognizing a reportable event and developing trust in reporting. The industries encouraged early reporting in a variety of ways. Overall, their experiences demonstrate that reporting is enhanced when (1) reportable events are broadly defined and allow reporting from a wide range of workers; (2) workers are able to describe the details of an incident or concern in their own words, with classification schemes applied by specialists at a higher level; and (3) both internal and external reporting options are available, along with some degree of report processing at the local level.

Broad Thresholds and Open Reporting Are Useful Features When Starting an SRS

In the three case industries, an early challenge was workers' lack of understanding of what should be reported. In each of the industries, the creation of an SRS involved broadening workers' concepts of safety events, in addition to accidents, that were worthy of reporting. Nevertheless, early reporting still tended toward accidents and technical issues—accidents because they were fairly evident and harder to hide and technical issues (as opposed to human factors) because the external nature of the fault provided some distance from individual blame. Reporting these technical events helped workers become more comfortable with reporting and provided objective links between their reports and systemic safety improvements, according to several industry officials. Over time, workers' ability to identify less concrete, but equally unsafe, nontechnical issues grew. The industries managed this growth, in part, by keeping the threshold and definitions for reportable events simple. In some cases, direct reporting—as opposed to reporting hierarchically, up the chain of command—was used to eliminate the fear that workers might have about reporting a mistake to the boss. Open reporting of events from several workers—especially those in different occupations—provided more raw data in the search for underlying causes, as well as information about the event from a variety of perspectives.

Aviation

The ASRS used a broad definition of reportable events and allowed all frontline aviation personnel to report them. Any actual or potential hazard to safe aviation operations are included in reportable events, thus expanding to areas on the risk pyramid beyond "accident." Serious accidents are not reported to the ASRS, since they are already covered by the NTSB. While reporting is available to all participants in the national aviation system, for several decades, the majority of reports were from pilots. After outreach and initiatives—such as revised specialized forms— the ASRS has in recent years seen modest increases in reports from diverse groups of workers, such as maintenance workers, enhancing the potential for analysis of single incidents from a variety of perspectives. To reduce the loss of information that could occur if reports from frontline workers are filtered through work hierarchies, the ASRS makes it possible for individual aviation workers to report directly to the central collection unit within NASA.

Commercial Nuclear Power

Individual nuclear plants operate corrective action reporting programs, which feed into INPO's SEE-IN system. The plant-level corrective action programs have a zero threshold for reporting—that is, workers can report anything of concern. To make the definition for reporting clear to workers, INPO characterizes the reporting threshold in terms of asking workers to report events that they would want to know about if the event had happened elsewhere.[46] In addition to establishing low reporting thresholds, a broad spectrum of workers are encouraged to report to the plant's corrective action programs. Open reporting and low reporting thresholds are necessary to ensure the fullest coverage of significant event reporting, according to an INPO liaison. While the individual plants are expected to assess and address the bulk of reports, they must also identify the most significant reports to send to INPO. Plants forward between 3,000 and 4,000 concerns to INPO each year from the estimated 400,000 concerns reported and resolved at the plant level through their corrective action programs. To ensure all staff are encouraged to report any event of interest, INPO examines the robustness of the plant's reporting culture during biannual plant inspections. As part of this process, INPO also compares corrective action reports to SEE-IN data to determine whether there are reports in the corrective action system that were not forwarded to INPO that should have been. If such discrepancies arise, these cases are discussed with plant managers to educate and clarify the plant's reporting thresholds to INPO.

Health Care

Prior to the SRS program, VA hospital workers were accustomed to reporting only the most serious events, such as inpatient suicides or wrong-site surgery. The VA SRS program expanded the definition of reportable events to include incidents—such as close calls or errors that caused no patient harm—in recognition of the value of incident data in detecting systemic safety problems.[47] Despite the conceptual shift in reporting expectations, in our 2004 report, we found that 75 percent of clinicians we surveyed at four facilities understood these new reporting requirements. In addition, the SRS program was designed to allow direct reporting from any member of the medical center staff to the patient safety manager. This expansion—beyond the previous expectation that nurses would report to their supervisors—was made in recognition of the power relationships among clinicians that might inhibit reporting. As a patient safety manager noted, the change in reporting expectations was evidenced when a chief surgeon came to report instances of mistaken patient identity in the surgery.

Encouraging Workers to Report Incidents in Their Own Words Facilitates Reporting Initially

In all three industries, delaying the launch of an SRS for development of a formal error classification scheme would have been unpalatable in light of significant pressure to implement solutions following serious events. Further, some safety experts believe rigid early classification of error can limit new knowledge and insights. In the absence of such schemes, the industries allowed reporters to give detailed narrative accounts of the incidents or concerns in their own words. As the industries' comfort with error terminology develops, some SRSs may encourage reporters to classify certain aspects of events in order to facilitate industrywide analyses.

Aviation

ASRS reports are primarily experiential narratives in the words of the reporters. Although the heavily regulated aviation industry had event definitions for rule enforcement, studies have concluded that the ASRS was begun without a formal classification of errors.[48] The unstructured nature of the narrative reports is an analytic challenge. However, the ASRS has developed a set of 1,200 separate codes that facilitate the analysis of aviation risk. Recent FAA activities are focused on the benefits of an integrated data system for safety events that combines ASRS's narrative reports and other reporting systems. Understandably, international aviation safety organizations have declared common reporting methods— including terms and forms—best practices.

Commercial Nuclear Power

The corrective action reporting programs at each plant collect information as narratives in the workers' own words. Corrective action reports are reviewed at the plant level by a team of managers and specialists. As part of this review, the team determines what actions, if any should be taken to address the issue, and reports are sorted and some level of classification is applied. Most corrective action reports are dealt with at the plant level. Only reports that rise to a defined level of significance—as determined through the review process—are sent on to INPO. While the reports sent to INPO do maintain narrative description of the event, they also classify specific aspects of the event. INPO further sorts and classifies these reports and produces various levels of industry alerts based on this review.

Health Care

According to a VA official, the SRS program was launched without an error classification system at the reporter level. Considering that even now the science for developing a formula for public reporting is evolving, he noted that the time it would have taken the VA to develop such a system would have delayed the launch by several years. Instead, the classification is done centrally. The VA has maintained this process because it believes that application of an error classification scheme is best done at higher levels by, for example, the patient safety managers. The VA official observed that the Agency for Healthcare Research and Quality (AHRQ) has been working on a set of error terms for nearly 5 years; however, there is, to date, no industrywide agreement on error or adverse event terminology in health care, although one for select health care institutions is under review.[49]

Reporting Options with Some Local-Level Processing Facilitates Reporting Initially

The initiation of SRS programs in two industries was driven by urgent circumstances, before there was time to assess workers' willingness to report. However, while program developers did not know everything about the problem, they did know that existing knowledge about the workforce culture could provide some basis for planning—that is, if employers suspect they have a mistrustful workforce, they can plan for it. In addition, the industries recognized that the value of local-level processing for improving safety culture and awarding responsibility for safety to the frontline was too great to completely give to an outside entity. Therefore, they developed a bilevel process for assessing safety data at both the local and industry levels.

Aviation

The airline industry manages the tension between trust and ownership in SRS reporting by offering a variety of internal and external, as well as local- and industry-level, reporting options. The ASRS (an external reporting option) was originally managed by the FAA, but within a year, it was moved to NASA—an honest broker—because of concerns that reporting directly to the regulator would discourage reporting. While separating the reporting function from regulation encouraged reporting, it may have fostered unconstructive perceptions of the ASRS among some FAA staff. Specifically, the 1994 NAPA evaluation found that FAA workers may not understand the ASRS and, consequently, devalue it. While the ASRS receives reports directly from reporters, the FAA's Voluntary Safety Programs branch (VSP) launched a bilevel SRS program in which 73 airlines are primarily responsible for receiving and processing reports and implementing solutions. By selecting a private structure for these SRSs, the FAA gets the entity closest to the local context to analyze reports and develop and implement solutions. A selection of the systemic problem reports is transmitted to the FAA's Aviation Safety Information Analysis and Sharing program, which the FAA uses to develop industrywide guidance and regulations to improve safety.[50] More than 60 percent of reports to the ASRS also appear in the other VSP's SRSs.

Commercial Nuclear Power

In the commercial nuclear power industry, most safety reports—an estimated 400,000 annually—are managed at the plant level, according to an INPO liaison. There is no confidentiality for individual reporters to their plant's SRS; instead, the reporting system relies on developing an open reporting culture. Each plant is responsible for sorting, analyzing, and implementing corrections for most of the reports to their corrective actions program. The reporter's identity is not revealed when the more serious events are sent on to INPO. INPO created a bilevel reporting structure because it lacked the resources to handle 400,000 reports annually and because it sought to involve the plants by giving them some ownership of the safety improvement system. However, recognizing the need for an industry-level assessment of safety data, INPO uses the more serious event reports from plants to develop industry alerts and safety recommendations.

Health Care

In the absence of specific information about workers' trust in reporting to an internal system, the VA could not be certain it had a safety culture that would support open local reporting. However, they knew nurses and pharmacists were "rule followers," while physicians had more discretion. The VA handled this uncertainty by initiating both internal and external reporting options. One reporting option, which emulated the ASRS model, was designed to enable workers to report directly to NASA—a contracted, external entity—confidentially. After operating both reporting options for nearly 10 years, the NASA-run system was discontinued for budgetary reasons at the end of fiscal year 2009. While the PSIS enables workers to report to an internal entity—the hospital's patient safety manager—the external NASA option provided more confidentiality and some measure of anonymity; the internal option provides personal contact and confidentiality, but not anonymity. Even with its much lower report volume—about a 1 to 1,000 ratio of reporting for the PSRS compared to the PSIS—for over 8 years, the system contracted to NASA provided a confidential alternative for workers who felt that they could not report to their own hospital, providing a safety valve or insurance policy of sorts. In addition to dual reporting options, the VA also planned for internal and external processing options. The NCPS intended that hospital-level report collection and processing—including root cause analysis and the development of systemic changes—be deliberately assigned to the individual hospitals to give workers on-the-job learning, and we found the experience drove home to clinicians that the SRS was a nonpunitive, solution-developing system. While reports are processed by a higher-level entity, the NCPS, to facilitate identification of issues with systemwide safety implications, local-level processing is also maintained because it provides a sense of ownership and immediacy in solving problems.

Lesson 3: Strong Legal Protections and Incentives Encourage Reporting and Help Prevent Confidentiality Breaches

Each industry we examined grappled with how to balance the regulatory tradition of punishing workers (or entities) for safety events with legal protections and incentives for reporting. Under most current laws, reports generated before an accident are considered discoverable evidence afterwards.[51] Such laws may deter companies from soliciting and collecting reports about safety problems and workers from reporting them. To address these concerns, the three industries offered a variety of mechanisms for protecting and encouraging reporting, including confidentiality provisions, process protections, and reporting incentives. Confidentiality provisions, rather than anonymous reporting, are the most common approach to protecting reporters' identities because they allow follow-up with the reporters; however, their protections are not ironclad. And, as SRS program managers in some of the industries discovered, even the perception that confidentiality can be, or has been, breached can discourage reporting. In the three industries, most of the laws supporting SRS confidentiality protections are a patchwork of older laws not originally intended to back up an SRS. Most also have exceptions to confidentiality if Congress or law enforcement agencies demand access to the protected documents. Some of the systems rely on existing laws, such as exceptions in the Freedom of Information Act (FOIA); other systems have a legal and regulatory basis crafted for related purposes. As SRS failures in other countries illustrate,[52] legal protections can be strengthened or weakened through legislative action.

Recognizing the fragility of confidentiality provisions, the three industries also relied on processes and incentives to protect and encourage reporting. Processes, such as deidentification of reports, support confidentiality provisions. Some industries apply it to both the reporter and the organization or unit involved. Data deidentification at the organizational level supports organizational buy-in for reporting, makes it less likely that reporters will be discouraged from reporting, and facilitates industrywide sharing by removing fear of reprisal. In addition, limited immunity provisions or small rewards were used, in some industries, as incentives to encourage safety reporting, especially in environments of mistrust. Limited immunity provisions apply when certain requirements— such as timely reporting—are met. These provisions provide reporters (individuals or organizations) with a means for avoiding or mitigating civil or regulatory penalties. With respect to rewards, even seemingly small incentives can be effective in promoting trust in reporting.

Aviation

The FAA protects its reporters through a combination confidentiality and limited immunity, relying on regulation, policy statements, and procedural or structural arrangements.

For the much older ASRS, confidentiality is maintained both as part of the interagency agreement between NASA and the FAA and through procedural efforts, such as deidentification of reports, as well as regulation. Section 91.25 of the Federal Aviation Regulations prohibit the FAA from using information obtained solely from these ASRS reports in enforcement actions against reporters unless criminal actions or accidents are involved. Specifically, after following up with the reporter and analyzing the report, the NASA office removes information that could identify the reporter, including the reporter's name, the facility, airline, or the airport. NASA destroys the identity portions of the original reports so that no legal demands could reveal them. The ASRS's information processing and deidentification of reports has ensured the confidentiality of its reports for over 30 years, despite pressures from the regulator and outside entities to reveal them. To strengthen the confidentiality agreement between the FAA and NASA, the FAA has determined by regulation that it will generally not use reports submitted to NASA in enforcement actions and provides some disciplinary immunity for pilots involved in errors.[53] In contrast, for several of the carrier-run SRSs initiated since 1997, reports are protected from legal enforcement action by the FAA only by policy.[54] However, despite the combined legal and procedural bases for protecting aviation SRS data—for both the ASRS and the other SRSs the FAA supports—there are pressures to violate SRS confidentiality. After recent judicial decisions forced disclosures from an SRS managed by the VSP branch, four major airlines withdrew from a voluntary program but have since rejoined.[55]

Commercial Nuclear Power

INPO operates under considerable confidentiality and maintains the ability to withstand legal challenges. Protecting the confidentiality of plants was central to the inception of INPO's safety efforts, according to industry officials. While guaranteeing its member utilities confidentiality similar to that in a doctor-patient relationship, INPO has also cultivated an open questioning attitude as the wellspring of safety reporting. While individual reporters receive no confidentiality, the reporting system relies on developing an open reporting culture. Under an INPO-NRC Memorandum of Agreement, reports and information that INPO makes available to the NRC will be treated as proprietary commercial information and will not be publicly disclosed.[56] INPO maintains legal resources for future confidentiality challenges. In INPO's bilevel system, reports sent to INPO do not identify the reporter, and INPO's confidentiality includes carefully guarding the identity of individual plants or utilities. For example, INPO does not reveal plants' safety scores. NRC officials reported that their process also guards against release of INPO information, such as looking at INPO's reports but not taking possession of them.[57]

Plants' interests in avoiding negative consequences also serve as an incentive for reporting. In particular, plants' fear of exclusion from INPO and interest in avoiding negative comparisons to other plants are tools the industry uses to promote reporting and workplace safety. An industry reality is that U.S. nuclear power plants are "hostages of each other," in that poor safety on the part of one plant could damage the entire industry's future.[58] In addition, the NRC and insurers would be made aware of a plant's exclusion from INPO, leading to increased insurance costs, as well as a loss of accreditation for training programs, which would result in more regulatory involvement by the NRC. The NRC and INPO identified other incentives that encourage nuclear plants in their current safety efforts, including (1) NRC credit on penalties if a plant identifies and corrects its own accident precursors, (2) the high cost of corrections, (3) the negative effect of safety events on stock values, (4) the loss of public confidence, and (5) insurance rates.

Health Care

The confidentiality of the SRS records that the VA hospital administration maintains is protected from disclosure by 38 U.S.C. § 5705—a law that predated the establishment of the SRS by over 15 years. This law prohibits the disclosure of records that are part of programs to improve the quality of health care. Sanctions, including monetary fines, are attached to disclosure violations, but there are exceptions to the confidentiality of the records, including demands by law enforcement agencies or Congress. More recently, the Patient Safety and Quality Improvement Act of 2005[59] provided similar confidentiality provisions, including fines for disclosure, for voluntarily submitted SRS-related documents from all U.S. hospitals.[60]

The bilevel structure of the VA's internal SRS facilitates deidentification. Individual hospitals collect and analyze reports and develop systemic fixes for their own hospital. Subsequently, the hospital sends reports and analyses—which are stripped of information that could identify individuals—to the central NCPS. The external, NASA-run SRS also deidentified reports. In addition, NASA destroyed the identification section of original reports in a process similar to that used for ASRS reports.

The VA does not grant immunity for intentionally unsafe acts or criminal behavior, nor does the safety program replace VA's existing accountability systems. However, individual facilities have used rewards as incentives, such as cafeteria coupons or cookies, to encourage reporting. In addition, hospital-level awards, such as awards to VA Medical Center directors from the NCPS, have also been used to encourage their support for reporting, analyzing selected reports in a timely way, and following up to mitigate risks identified in their reports and analyses.

Lesson 4: A Central, Industry-Level Entity Facilitates Lesson-Sharing and Evaluation

While some of the SRSs in the three industries have local-level processes for analyzing safety reports, they also have a central, industry-level entity that collects, analyzes, and disseminates safety data and makes recommendations. These industry-level entities facilitate feedback and evaluation by (1) elevating facility-level safety data to industrywide lessons and disseminating them across the industry, including internationally, and (2) assessing safety culture and identifying units or worker subgroups in need of outreach or intervention.

Some industry SRSs offer direct reporting to a central, industry-level entity, which is responsible for processing, analysis, and dissemination. For others, reporting takes place at the local level. While some level of report processing, analysis, and dissemination takes place at these local facilities, full or deidentified safety data are sent to a central, industry-level entity. Sending reports up to a central entity ensures that safety fixes identified through local processes are not lost to the rest of the industry. At the same time, local analysis and feedback can demonstrate the system's value to workers and reinforce reporting. Because the central entity receives safety data from multiple organizations—whether through direct reporting or from local-level systems—the volume and variety of information increase the potential for identifying systemic issues and improving safety industrywide. In addition, the industries recognize that a central, industry-level entity might be necessary for bringing some difficult safety problems to light. This is because the central entity is more likely to consider the interests of the industry, whereas local-level managers might resist identifying systemic issues that would put personal or organizational interests at risk. These central entities, because of their position as industry representatives, are also in a better position to disseminate lessons across the industry and internationally. They provide a single source for industrywide notices of varying urgency, regular online newsletters, policy changes, briefings, and data systems. In addition, some of these entities have staff with internationally recognized safety experts— expertise which has been leveraged worldwide to inform international safety recommendations and SRS design.

The central, industry-level entities are also in a better position to facilitate evaluation, including safety culture assessment; identification of reporting gaps (access to safety data from across the industry offers the potential for analysis of gaps across particular locations, organizations, or occupations); and needed system modifications. Furthermore, such entities often have access to other safety data, such as inspection information. This information can be compared with reporting data in order to identify sites in need of outreach and training. Such systemwide visibility provides an ideal position from which to conduct SRS evaluations. Industry experts we spoke with believe that their industries are safer, in part, as a result of their SRS programs. In limited cases, the central entities have been able to conduct evaluations or use performance metrics to assess safety culture improvements and the role of the SRS in those efforts, as is recommended under the Government Performance and Results Act.

Aviation

The ASRS shares lessons with all levels of the domestic aviation community and has served as a model of aviation safety reporting worldwide. NASA's ASRS issues a series of industrywide notices based on ASRS reports, which are graded on the basis of the urgency and importance of identified safety issues, and it has been recognized worldwide as a model for collecting data from frontline workers. NASA provides "alerting" messages to the FAA and the airlines on safety issues that require immediate attention. NASA also disseminates ASRS information via a monthly online bulletin, *CALLBACK*, to 85,000 members of the aviation community on safety topics such as summaries of research that have been conducted on ASRS data. Unions and airlines use this information in safety training. Among the SRSs we are aware of, only the ASRS offers access to its event database for outside researchers to conduct analysis and for ASRS staff to perform specially requested analyses for the FAA, NTSB, and others. The FAA also maintains an industry-level office—the VSP branch—which oversees seven different voluntary reporting systems, including the ASRS. Data from these SRSs provide information on events that would otherwise be unknown to FAA or others, and VSP's role is to facilitate sharing of these data at the airline and industry levels. We observed VSP and ASRS staff representing U.S. airline safety interests at an international aviation safety reporting meeting to share lessons on aviation safety and SRS design and implementation. Such participation offers opportunities for safety improvement in aviation worldwide. For example, VSP and ASRS staff have supported efforts to develop safety reporting systems worldwide because aviation safety does not stop at the U.S. border. Most foreign aviation SRSs have been based on the ASRS model. The international aviation safety organization, the International Civil Aviation Organization, has called for each country to have an independent aviation safety reporting system similar to ASRS.

Despite the benefits of these SRSs, formal evaluation has provided insights for system improvement. For example, the FAA requested the NAPA evaluation of ASRS, which recommended the ASRS modernize by implementing actions, such as collecting and disseminating reports in electronic formats to better meet the needs of the aviation community.61 Currently, ASRS safety reports and monthly newsletters are primarily transmitted by e-mail. In addition to ASRS-specific evaluations, the FAA has access to more investigations of aviation safety culture conducted over the last decade. For example, special studies of aviation specialists, such as controllers and maintenance workers, have identified reasons for their lower reporting rates. These studies revealed specific aspects of cultures in these professions that would discourage reporting. For example, controllers were highly focused on bureaucratic boundaries that enabled them to define away—rather than report—unsafe conditions they perceived to be outside their responsibility. Alternatively, according to FAA officials, they found a strongly punitive culture among maintenance workers that led workers to assume that if a supervisor told them to violate a rule, it did not create an unsafe—and hence reportable— condition. These studies made possible targeted efforts, such as a reporting program just for controllers, that resulted in a growing proportion of safety reports from nonpilots.

Commercial Nuclear Power

INPO's lesson-sharing program uses the Nuclear Network—an industry intranet—for sharing safety information. This network houses event data that plants can access and is a platform for INPO to disseminate alerts. Information transmitted via this system includes Significant Operating Event Reports—the highest-level alert document—as well as experiential and nuclear technical information. Plants can also use the network to ask questions or make comments that can be sent to one, several, or all users. Apart from the direct feedback reporters receive from the plant, the key to getting workers to participate in reporting was through seeing—via the Nuclear Network—the corrective actions developed in response to reports they had made, according to the INPO liaison. INPO is seen as a model for other national and supranational nuclear safety organizations, such as the World Association of Nuclear Operators, an organization representing the global nuclear community. As such, INPO has recently begun to participate in the Convention on Nuclear Safety, a triannual international commercial nuclear safety effort.[62]

INPO also evaluates plants' safety improvement programs, although the evaluations are generally not publicly available, according to an INPO liaison. INPO performs a type of "gap analysis" at the biannual on-site plant inspections and conducts safety culture surveys with a sample of staff before each.[63] Reporting gaps are evaluated at the plant level (not by occupation or work group) by looking for reductions in report volume and mining the plant's corrective action reports. A reduction in reporting year to year is interpreted as an indicator of a potential problem rather than an improvement in safety conditions, because such reductions can indicate a lack of management support for reporting. In addition, if a plant receives a low safety score as a result of inspection findings, INPO provides extra attention and assistance by assigning a team of industry experts to engage in weekly consultations with plant directors, review corrective actions, discuss plant needs, develop solutions, and provide peer assistance and accompaniment to seminars.

Health Care

In its position as the industry-level entity responsible for the VA SRS, NCPS creates and disseminates key policy changes to the VA health care system in response to trends identified from patient safety reports. For example, the NCPS (1) designed and implemented a program that promotes checklist-driven pre- and postsurgical briefings that, according to the SRS program director, have been associated with reduced surgical mortality across the VA hospital system and (2) developed new requirements for CO2 detectors on every crash cart for checking safe intubations outside of operating room settings. The NCPS has played a role in disseminating its SRS model and tools for safety improvement to other U.S. states and federal agencies, including the AHRQ. Specifically, the NCPS provided training to all 50 states and the District of Columbia via the Patient Safety Improvement Corps, a program funded by the AHRQ.[64]

The VA -supplied state training contributed heavily toward building a common national infrastructure to support implementation of effective patient safety practices.[65] Further, after attending the VA seminars, several foreign countries implementing their own SRSs have adopted tools developed by the VA.

The NCPS has also conducted evaluations of the SRS program, which have provided information for SRS and safety culture improvements. For example, in 2008, the NCPS published a study of the effectiveness of actions hospitals developed in response to SRS reports of adverse drug events.[66] They found that changes in clinical care at the bedside—such as double- checking high-risk medications—and improvements to computers and equipment were effective solutions, but training was not. In addition NCPS has conducted three safety culture surveys, the most recent of which enabled identification of safety culture differences among staff subgroups in order to target outreach and training. To support future evaluations of this kind, the NCPS established several criteria to assess the quality of local-level processes for reporting, analysis, and safety improvement.

The CDC and APHIS Have Taken Steps to Improve the Usefulness of the TLR Reporting System; Lessons from the Literature and Case Studies Suggest Additional Steps

The CDC and APHIS Select Agent Program (SAP) has taken steps to improve reporting and enhance the usefulness of the theft, loss, and release (TLR) reporting system as a safety tool.[67] Additional steps to improve the TLR system, as suggested by the literature and case studies, include increased awareness of the culture in biological labs and improvements in the three key areas—reporting and analysis, protections and incentives, and feedback mechanisms. See appendix II for a summary of lessons derived from the literature and case studies that can be applied to the TLR system.

The CDC and APHIS Recognize the TLR Reporting System's Usefulness as a Safety Tool; Lessons Indicate That Increased Awareness of Labs' Culture Could Enable Targeted Outreach and Training

Recognizing the usefulness of the TLR system as a safety tool, the CDC and APHIS SAP has dedicated resources to manage the system. The TLR reporting system for select agents was developed in 2002, after the 2001 anthrax attacks.[68] As the number and types of reported incidents increased, an outcome of the new reporting requirements, the agencies implemented processes to utilize the TLR system as a tool to manage the Select Agent Program. In addition, the CDC reassessed its administration of the system to consider how it could be used as a safety tool, rather than just a recording system. To its credit, the CDC employed a safety science expert to manage the TLR reporting system and is now exploring ways of using the TLR data to identify systemic safety issues. APHIS has also utilized the TLR as a tool to identify trends such as (1) gaps in administrative oversight of personnel and training and (2) weaknesses in safety and security policies and procedures in regulated entities. Each TLR is reviewed by a compliance officer, security manager, and subject matter experts to identify trends and areas of concern. Identified issues are subsequently discussed with the reporting facility's senior management, with additional monitoring and inspections as needed.

The CDC and APHIS also rely on periodic on-site lab inspections to get an understanding of the culture, with respect to safety and reporting, and identify areas for outreach and training. The agencies inspect labs to ensure that they are in compliance with the safety, security, training, and record-keeping provisions outlined in the regulations. As part of this process, the agencies use checklists developed from regulations and nationally recognized safety standards to review laboratory safety and security and to develop observations. In addition, the agencies interview lab staff and examine documentation, such as medical surveillance documents, exposure or incident records, and minutes from Institutional Biosafety Committee meetings.

Review of such documentation can provide an indication of possible incidents with select agents or toxins. During these inspections, the CDC and APHIS officials seek to (1) identify gaps in knowledge about safety and reporting and (2) report on areas needing improvement.

The information the agencies derive from these inspections and from TLR reports can provide useful information about the culture of safety and reporting within labs. However, lessons from the literature also suggest that systematic assessment of the culture, such as through ongoing surveys or studies, can provide invaluable information about how the specific working environment can affect perceptions of safety and reporting requirements.[69] These perceptions—and variations, for example, within or across working environments or occupations—can affect what is considered a reportable event; feelings of responsibility for or fear of reporting; and the value of reporting safety events. For example, studies examining the effects of culture on safety and reporting in the aviation and health care industries have found that perceived occupational hierarchies, such as between doctors and nurses or pilots and cabin crew;[70] authority structures;[71] organizational factors;[72] concepts of justice;[73] and other factors can affect safety and reporting.

According to CDC and APHIS officials, they have no plans to arrive at such an awareness through cultural assessment. Nevertheless, agency officials agree that culture matters when it comes to safety and reporting. For example, they noted that culture may differ by a lab's size and level of resources. Larger labs or labs with more resources tend to have better safety and reporting. Other agency officials noted that, based on career experiences, they have become aware of safety differences across different types or levels of labs. According to a CDC official, staff in higher-level labs, such as BSL-4 labs, have recognized the danger of the material they are working with. These facilities are also more likely to have biosafety officers, whose presence, according to the CDC official, tends to make workers more conscientious about safety. Another official noted that, while you might find sandwiches or soda in the refrigerator of a BSL-2 lab, these items would never be found in BSL-4 labs. Safety culture differences between clinical and research labs were also noted by CDC officials. Such variation in culture across labs was also noted by domestic and international biosafety specialists we spoke with. Despite recognition of such variation across labs, officials stated, the CDC does not have a unified position on the issue, and the research does not exist to definitively establish safety culture differences by lab type, occupation, or sector. Greater awareness of cultural influences and how they affect safety and reporting in the labs could (1) help the agencies better target outreach and training efforts and (2) provide insights into whether reporting system design and implementation changes are needed to address lab variations in safety and reporting.

The CDC and APHIS Have Taken Steps to Better Define Reportable Events; Lessons Indicate That a Broadened Definition Could Further Enhance Collection of Safety Data

The CDC and APHIS SAP has taken steps to better define reportable events, which can increase the likelihood that workers will report when required. For example, in early 2008, the CDC and APHIS published the *Select Agents and Toxins Theft, Loss and Release Information Document,*[74] which includes detailed scenarios on what and when to report. Since the TLR reporting program was established in 2002, the agencies have seen reports increase substantially; since a 2008 initiative to better inform the lab community of incident-reporting requirements, the CDC and APHIS noted that they receive approximately 130 incident reports per year. The types of labs reporting have also broadened. According to the CDC, the increased reporting is the result of better awareness of and compliance with reporting requirements, rather than an increase in thefts, losses, or releases.[75] Indeed, of the reported TLRs, there have been no confirmed thefts, one loss, and only eight confirmed releases.

To clarify reportable events, the Select Agent Regulations require that the individual or entity immediately notify the CDC or APHIS upon discovery of a release of an agent or toxin causing occupational exposure, or release of a select agent or toxin outside of the primary barriers of the biocontainment area. The agencies' *Select Agents and Toxins Theft, Loss and Release Information Document* further clarifies reportable events. The document defines a release as a discharge of a select agent or toxin outside the primary containment barrier due to a failure in the containment system, an accidental spill, occupational exposure, or a theft. Furthermore, any incident that results in the activation of medical surveillance or treatment should also be reported as a release. The document also emphasizes that occupational exposure includes any event in which a person in a registered facility or lab is not appropriately protected in the presence of an agent or toxin.[76] For example, a sharp injury from a needle being used in select agent or toxin work would be considered an occupational exposure. While these reporting requirements are fairly broad, they do require a degree of certainty about the occurrence of an event. But, in some cases, recognition of a reportable event may come only when consequences are realized.

While the agencies' steps to better define reportable events can increase the likelihood that recognized events will be reported, according to the literature and biosafety specialists, lab workers are often unaware that a release has occurred unless or until they become sick. For example, early studies of LAIs found that as many as 80 percent of all reported LAIs could not be traced back to a particular lab incident. A more recent study found similar results.[77] The absence of clear evidence of the means of transmission in most documented LAIs highlights the importance of being able to recognize potential hazards because the likely cause of these LAIs is often unobserved. While a great deal is known about micro-organisms to support safe lab practices, microbiology is a dynamic and evolving field. New infectious agents have emerged, and work with these agents has expanded. In addition, while technological improvements have enhanced safety, they can also introduce new safety challenges. For example, failures in a lab system designed to filter aerosols led to a recent company recall of this system.[78] The dynamic nature of the field, coupled with the difficulty of identifying causal incidents in LAIs, suggests substantial potential for unintentional under-reporting. In such an environment— where workers are waiting for an obvious event to occur before reporting—a significant amount of important, reportable safety information could be lost. Consequently, while reporting requirements for releases may now be clear for many incidents or for observed consequences, broader reporting thresholds may be necessary to accommodate emerging safety issues and the unobserved nature of many LAI events.

According to lessons from the literature and case studies, expanding reporting thresholds—in this case, to include observed or suspected hazards—can help capture valuable information for accident prevention. The industries in the case studies all struggled with how to recognize, and thus report, such events. However, over time, the feedback they received from these reports, in the form of specific safety improvements, helped workers develop familiarity and comfort with recognizing and reporting such events. An example in the lab community might be the practice of mouth pipetting, drawing an agent into a pipette by sucking on one end. At one time, mouth pipetting was a common practice, despite the high risk of exposure. Even though not every instance resulted in exposure or an LAI, some did, and eventually the activity was recognized as a potential hazard—an accident precursor. Expanding the TLR reporting threshold to include hazards could provide additional data that might be useful for safety improvement efforts. For example, INPO encourages reporting of events at all levels of the risk pyramid—including the hazard level—for the corrective actions reporting programs of nuclear power plants. This level of reporting ensures as complete coverage as possible of potential safety issues. For the TLR, reporting at this level could be voluntary or mandatory. Moreover, until a labwide voluntary reporting system is implemented, reporting at this level could further develop the reporting culture among select agent labs.

The CDC and APHIS Have Taken Steps to Protect Confidentiality, Which Can Encourage Reporting; Lessons Indicate That Limited Immunity Could Further Encourage Reporting

The CDC and APHIS SAP has taken steps to incorporate deidentification measures to further protect the confidentiality of entities reporting thefts, losses, or releases. While entity-specific information is protected from release under FOIA,[79] there was an instance when specific entity information was somehow leaked to the media after the CDC provided the data in response to a congressional request. As a result, the agency provides only deidentified report forms in response to congressional requests. In addition, to further support reporter confidentiality in the event of audit or congressional requests to view TLR information, the CDC has established an access-controlled reading room for viewing these reports. It expects these measures to prevent any future prohibited disclosure of entity-specific data, while special-need access to information about thefts, losses, or releases is provided.[80] According to lessons from the literature and case studies, even the perception of a confidentiality breach can quash reporting. Consequently, the agencies' measures to ensure confidentiality can increase confidence in reporting.

Apart from the requirement to report, labs also have some incentive for reporting. One such incentive, according to CDC officials, is labs' interest in avoiding increased oversight.[81] In addition, lab officials know that (1) select agents are on the list because they are dangerous and (2) it is of critical importance to promptly report incidents to ensure proper care of workers and the public. CDC officials stated, however, that too much discretion about what and when to report could result in the underreporting of more serious events. As the experiences of the case industries illustrate, protection of reporter confidentiality is an ongoing effort, even when strong legislative provisions exist to protect reporters' identities. Because, as mentioned above, even the perception of a confidentiality breach can quash reporting, strong incentives for reporting—such as limited immunity provisions—can balance these fears and encourage continued reporting, according to lessons from the literature and case studies.

If the CDC or APHIS discovers possible violations of the select agent regulations, the following types of enforcement actions may occur: (1) administrative actions, including denial of application or suspension or revocation of certificate of registration, (2) civil money penalties or criminal enforcement, and (3) referral to the Department of Justice for further investigation or prosecution.82 Currently, even if entities report violations, there are no provisions for receiving immunity from these enforcement actions. In the aviation industry, pilots face the possibility of similar enforcement actions for violations of regulations. However, the FAA provides some disciplinary immunity for pilots reporting violations of regulations to ASRS.83 Such immunity is in recognition of the fact that (1) information about pilots' errors is essential for identification of systemic problems and (2) pilots would be unlikely to report their errors without some incentive to do so. Similar provisions for limited immunity from administrative action or reduced monetary penalty could be offered to labs for some violations of select agent regulations. Although the CDC and APHIS have not yet explored this option, such an incentive could be a powerful tool for ensuring reporting compliance.

The CDC and APHIS are Uniquely Positioned to Support Data Sharing and Feedback Efforts, Including Evaluation

The CDC and APHIS are uniquely positioned to support feedback and evaluation efforts that are based on TLR information. The agencies' oversight responsibilities for registered labs and their recognized expertise in laboratory safety practices provides them visibility and authority across the lab community. Such a position, according to lessons from the literature and case studies, is ideal for (1) disseminating feedback from SRSs and (2) evaluating the effectiveness of the reporting program. Currently, the agencies have a process for providing feedback to the reporting institution, and are beginning to explore avenues for sharing safety lessons across the labs and internationally.

In addition, the CDC has begun using the data to develop lessons learned from reported information. Although deidentified reports are not available to the general public, they are being used for special research studies sponsored by the Select Agent Program. For example, information from deidentified reports has been used for conferences such as the yearly Select Agent Workshops, sponsored by the CDC, APHIS, and the Federal Bureau of Investigation. The agencies are also analyzing data on select agent release reports and plan to publish the findings in a publicly available, peer-reviewed journal.

Such feedback demonstrates the value of reporting, according to lessons from the literature and case studies. Lessons from the case studies also indicate that using SRS data to develop guidance and sharing such information internationally can support industrywide safety improvement efforts. For example, TLR data could provide valuable information for updates to the BMBL and World Health Organization guidelines, which can benefit the worldwide lab community.

When a lab reports a TLR, the CDC or APHIS provides feedback and, if necessary, follows up to determine the root cause or initiate surveillance. While the CDC recognizes the usefulness of TLR reports for generating data that can (1) help spot trends, (2) highlight areas for performance improvement, and (3) show limitations in current procedures, it is just beginning to collect enough data to see patterns of nonreporting, according to CDC officials. The CDC expects that in the future, it will have collected enough data, including inspection data, to identify reporting patterns and conduct targeted outreach to nonreporting labs.

However, the agencies do not yet have a specific plan to identify reporting gaps in order to develop targeted outreach and training or to assess the system's effectiveness. To further support targeted outreach, as well as system modification, evaluation is needed.

As we have previously reported, such evaluation can be a potentially critical source of information for assessing the effectiveness of strategies and the implementation of programs.[84] Evaluation can also help ensure that goals are reasonable, strategies for achieving goals are effective, and corrective actions are taken in program implementation.

For example, an evaluation of the ASRS program revealed the need to improve the usefulness of the system through system modifications and increased outreach to certain populations. According to CDC Select Agent Program officials, they have had general reviews, such as an HHS Office of Inspector General review and a federally funded, third-party review of procedures conducted by Homeland Security. However, these reviews did not focus on the effectiveness of the TLR reporting system.

EXISTING INFORMATION ON BIOLOGICAL LABS AND LESSONS FROM THE LITERATURE AND CASE STUDIES SUGGEST SPECIFIC SRS DESIGN AND IMPLEMENTATION CONSIDERATIONS

Safety reporting system evaluation literature and case studies of SRSs in three U.S. industries—aviation, commercial nuclear power, and health care—provide lessons for design and implementation considerations for a national biological lab SRS.[85] First among these lessons is the need to set system goals and assess organizational culture, as illustrated in figure 4. However, assessment of organizational culture is difficult in the context of U.S. biological labs because there is an unknown number of labs and, except for labs in the Select Agent Program, no entity is responsible for overseeing all labs. While many federal agencies have labs and are involved in the industry, no single regulatory body has the clear responsibility or directive for the safety of all laboratories.[86] Consequently, an important part of the goal-setting and assessment process for a biological lab SRS is determining the scope of labs to which the system would apply. For example, specific system goals, such as the ability to identify trends or incidence rates, may be possible with one type or level of lab, but not another. Similarly, assessment may reveal that differences in organizational cultures across lab types is so significant that appropriate SRS features for one type of lab would not apply well to another. Consequently, the scope of labs to which an SRS might apply could be addressed as part of the goal-setting and assessment process.

Until such a goal-setting and assessment process is completed, design and implementation options in the three key areas—reporting and analysis, reporter protections and incentives, and feedback mechanisms—can be considered in the context of available information on organizational culture in biological labs and potential goals for a biological lab SRS. In particular, the following can provide some context to guide early decisions for the design and implementation of an SRS for the lab community: biosafety research, experiences with the TLR reporting system and biosafety specialists' perspectives. Such context can be further refined once assessment and stakeholder input are obtained. In addition, the NIH has begun developing a prototype reporting system for a subset of its intramural research labs. Lessons from how this prototype system works for a subset of labs could also inform design and implementation considerations for a national biological lab reporting system.

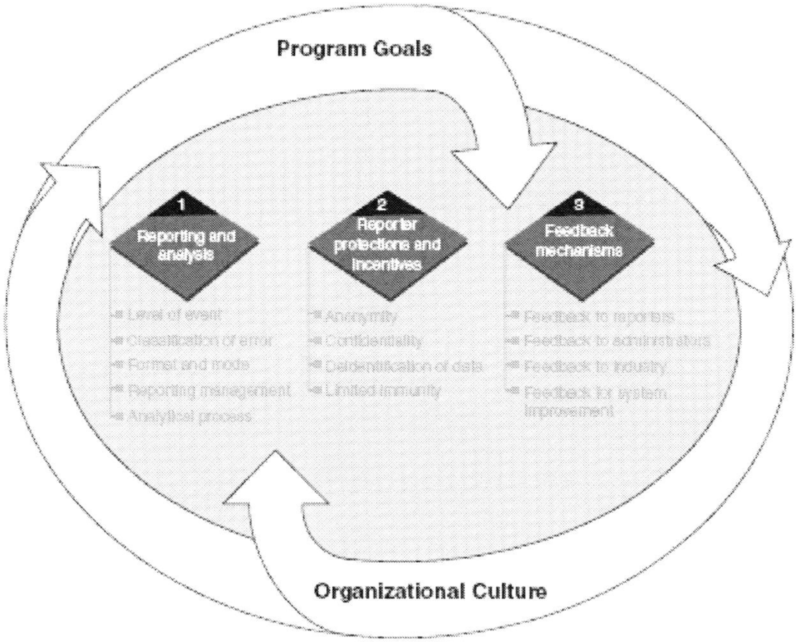

Source: GAO analysis of SRS evaluation literature.

Figure 4. Relationship of Program Goals, Organizational Culture, and the Three Key Areas.

In the Context of Existing Information, Lessons Suggest Several Features for Reporting and Analysis

Existing information about the potential goals for a biological lab SRS and the organizational culture of these labs suggest certain design and implementation features in the first key area: reporting and analysis. Figure 5 shows the relationship of program goals and organizational culture to this key area.

Level of Event, Learning Goal, and Culture Suggest Voluntary Reporting

The level of event of interest, probable SRS goals, and organizational culture all suggest voluntary reporting for a biological lab SRS. While the TLR reporting system for select agents is focused on incidents or accidents that pose the greatest danger to workers and the public, an SRS for nonselect agents could be used to gather information on hazards and potentially less serious incidents and accidents in order to collect precursor data. Systems that focus on less serious events and that collect precursor data to support learning rather than enforcement goals are generally associated with voluntary reporting, according to lessons learned. Voluntary reporting for a biological lab SRS also corresponds with the views of biosafety specialists we spoke with.

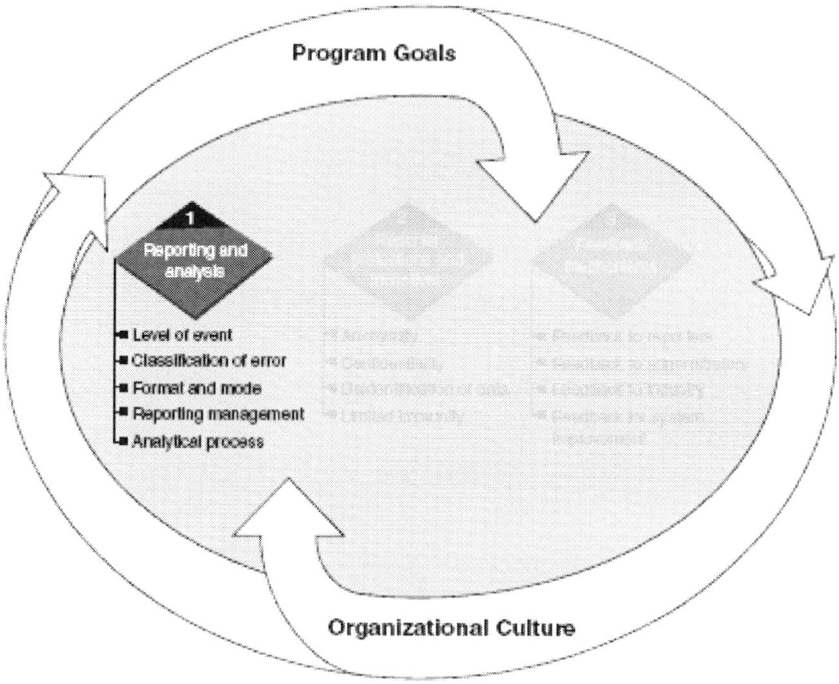

Source: GAO analysis of SRS evaluation literature.

Figure 5. First Key Area—Reporting and Analysis.

Laboratory Community's Limited Experience with Reporting to an SRS Suggests an Initially Open Classification Scheme

Reporting to an SRS—especially for incidents beyond LAIs or the theft, loss, or release of select agents—would be relatively new to the lab community. And although select agent labs have become familiar with reporting theft, loss, or release incidents, previous reporting failures indicate that, even among this subset of labs, reportable events may still be unclear. In such situations, allowing workers to report events in their own words, rather than asking them to classify the event as a certain type of hazard or error in order to report, can facilitate reporting. Classifying events—that is, applying standardized descriptions of accidents, incidents, and hazards—can facilitate safety improvement across the industry by providing a common language for understanding safety events. But classification can also limit reporting if workers are unsure of how to apply it. One solution for industries new to SRS reporting is to apply classification at a higher level, for example, through the event review or analysis process.

Ensuring the reporting process is as clear and simple as possible is especially important for the lab community. Although LAIs are widely recognized as under-reported, there is, at least, a long history of reporting these events among lab workers. However, lab workers do not have as much experience reporting events without an obvious outcome, such as an LAI. Many of the biosafety specialists we spoke with had difficulty envisioning the types of events—apart from LAIs—that might be reportable. In addition, even when LAIs do occur, many are never linked with a specific causative incident, so information about potential event precursors is never communicated or is difficult to identify. Difficulty recognizing exposure is a reality of work in these labs. LAIs often occur through aerosol exposure, and the activities that can create such conditions are numerous. However, all three case-study industries grappled with similar difficulties in recognizing and reporting events that did not result in obviously negative outcomes. One way the industries addressed this difficulty was to allow workers to report a broad range of events in their own words. Over time, as workers saw concrete results from their reports, such as improved processes or guidance, their ability to identify less concrete, but equally unsafe hazards and incidents—even those without obvious consequences—grew. Expecting lab workers to classify events in order to report them would likely limit reporting. In such situations, lessons learned suggest allowing workers to report events in their own words to facilitate reporting.

Diversity of Lab Community and Uncertainty about Reporting Population Suggest Multimode and Open Format Reporting Options, with Direct and Open Reporting

The lab community is organizationally diverse and the population of labs is unknown. Opening reporting to all workers, and offering multiple reporting modes (e.g., Web and postal), and using forms with open-question formats that allow workers to report events in their own words can facilitate reporting in the face of such uncertainty, according to lessons from the literature and case studies. Biological labs operate across a wide range of employment sectors, locations, and levels of containment. There are BSL-2, 3, and 4 labs in private, academic, and public settings across the United States. Staffing models for these labs are likely as different as the lab populations. Safety culture and reporting proclivity also vary across lab types. For example, according to biosafety specialists, clinical and academic labs—in contrast to government and private labs— face greater challenges to creating a safety culture and reporting events. According to one biosafety specialist, in academic labs, students expected to complete lab work before they have received adequate safety training may not feel they are in a position to demand such training. Specialists also indicate that higher-level labs (BSL-3 and 4)—especially the larger ones with better resources—have personnel, equipment, and/or processes to better support safety culture than lower-level, smaller labs with fewer resources. Furthermore, the consequences of accidents are so great at higher-level labs that the culture is generally more cautious. At lower-level labs, the perception of risk and actual risk are lower, so practices are not as stringent as they would be at higher-level ones.

The work environment at biological labs also varies. In particular, some work is done in teams and some individually, and some is completed overnight because of time-sensitive experiments in the research. In addition, the solo nature of much lab research means that a single lab worker may be the only one who knows about an incident. For lab work, the external visibility of accidents and incidents present in aviation or some areas of health care may not exist. Bioresearch errors are also a lot harder to spot than errors in other industries. For example, nuclear safety officers can use radiation detectors to determine whether breaches of protocol have occurred by identifying hot spots in suspicious areas, such as a phone outside the lab. No similar tracking mechanism exists for bioresearch. Therefore, the only objective proof of most accidents is that someone became ill. In addition, lab workers have little incentive to report if the incident occurred as a result of their own error, according to biosafety specialists. Although one specialist believes there is a fair degree of reporting on equipment failures because researchers generally want to ensure that the equipment is fixed.

Such variation has consequences for reporting. According to lessons from the literature and case studies, assessments can provide information about aspects of organizational cultures, structures, or processes that can affect reporting. However, a comprehensive assessment of this sort is difficult because (1) the population of labs is unknown and (2) no entity is responsible for conducting such an assessment. Given the uncertainty about cultural influences that may affect reporting behavior, more inclusive reporting options can facilitate reporting, according to lessons from the literature and case studies. For example, uncertainty about lab workers' access to reporting forms or ability to complete detailed forms can be minimized if (1) workers can report in whichever mode is most accessible to them (Web or postal) and (2) the forms do not require overly detailed or technical explanations.

In an environment where much of the work is done alone and incentives may not exist for reporting, an SRS that is open to all lab workers (including security and janitorial staff) can facilitate reporting where none might occur. Accepting reports from workers not directly involved in research can increase the volume of safety data that can be obtained. Multimode and open-reporting formats, as suggested above, support open reporting since staff with varying knowledge of biosafety terms—such as janitorial, security, or animal care staff—are still able to report incidents or hazards in their own words in the way that is most convenient to them.

Historically, the preferred model of biosafety reporting is hierarchical. This ensures that workers receive timely medical intervention and surveillance. Although it is important that workers have a mechanism for receiving immediate medical attention and surveillance when needed, a lot of important safety information could be lost if only supervisors or managers are allowed to report. Hierarchical reporting structures may limit the amount of useful safety data that can be received because a filtering process takes place at each level in the reporting hierarchy. As the information moves up the reporting structure, each person assesses whether the event is reportable. If the person decides that it is, he or she will report his or her own interpretation of events. Allowing all workers to directly report to an SRS removes this filter and can increase the number of reports and the amount of information collected from reports. For example, reports from multiple sources can enable analysis of events from multiple perspectives. While workers should always be encouraged to report potential exposures and other hazards to their supervisors so that they can receive timely medical attention, they should also be able to report incidents directly to an SRS.

Advantages and Disadvantages Inherent in Industry-Level and Local-Level SRS Administration Suggest a Dual Reporting Option

The HHS and USDA—as central, recognized authorities in the biological lab community—represent the kind of industry-level entities that, according to lessons learned, are necessary for effective dissemination and evaluation activities. However, the agencies' regulatory role in the Select Agent Program could inhibit voluntary reporting, suggesting that an alternative reporting mechanism may be necessary. According to lessons from the case studies, dual reporting options can facilitate reporting in such situations. For example, if workers are concerned about reporting safety events—either to an internally managed SRS or to the regulator—an external, independently managed SRS can be useful. Alternatively, if workers are comfortable reporting to a local SRS, these programs can be very effective when the information from local systems is fed to a central, industry-level entity that can analyze data across the industry and disseminate safety improvements industrywide.

While each case study industry differs in its approach, all three rely on dual (or multiple) reporting options. Specifically, the FAA relies on the independently run ASRS, as well as seven other key reporting programs, to collect safety data. Events that meet reporting requirements can be reported to the ASRS—meeting the need for an independent reporting mechanism for those concerned about reporting to either their local (airline-run) SRSs or to the regulator. In addition, as part of the FAA's other reporting programs, the FAA receives SRS data from the airlines, which they use to develop industrywide safety improvements. The commercial nuclear power industry also has reporting options. While each plant has a reporting system for corrective actions, a portion of the more significant reports are passed on to INPO for development of industrywide safety improvements. Individuals and plants also have the option to report to NRC's Allegation Program. Finally, in designing its reporting program, the VA created two reporting options—one externally managed by NASA and one local, hospital-based program in which safety data are sent on to VA's National Center for Patient Safety (NCPS) for development of industrywide safety improvements. While the industries might encourage workers to use one option over another, they are still able to report to the system most comfortable for them. Both options, however, utilize an entity with industrywide visibility and recognized authority to disseminate SRS information and direct system evaluations.

An external, independently managed SRS for the lab community offers several advantages, including the (1) potential to reduce workers' fear of being punished for reporting, (2) ability to contract for system management, and (3) centralization of safety data. Nevertheless, since the individual labs have the most intimate knowledge of staff, pathogens, and operations, several biosafety specialists adamantly indicated that the lab facility was the appropriate level for reporting and analysis. According to lessons from the literature, as well as the perspectives of biosafety specialists, analysis of safety reports should be done by qualified biosafety professionals and others with appropriate expertise or knowledge. In addition, processes for local-level collection and analysis of SRS reports can facilitate worker buy-in for reporting, according to lessons from the case studies. However, not all labs have the same resources for collecting and analyzing reports. Furthermore, the focus on safety culture across the lab community may not be sufficient to support an SRS program that operates only at the local level. But local-level support—as well as encouragement of reporting, receptivity to safety concerns, and regard for the field of biosafety—is central to a robust reporting program. Even if there is receptivity to biosafety issues, when safety is the responsibility of those internal to the organization, there may be conflicts of interest in addressing safety issues. While safety improvements are most useful when shared across the lab community, sharing this information may raise institutional concerns about funding streams, public perception of the institution, and professional standing of lab workers, according to biosafety specialists we spoke with.

Given the advantages and disadvantages of SRS administration at both the local and agency levels, dual reporting options may be necessary, at least initially. For example, the VA initiated its safety reporting program with both internal and external options. Although the VA canceled the NASA-run program after nearly 10 years, in recognition of the importance of an external reporting option, some efforts to reestablish the system continue.

In the Context of Existing Information, Lessons Suggest Several Features for Reporter Protections and Incentives

Existing information about the potential goals for a biological lab SRS and the organizational culture of these labs suggest certain design and implementation features in the second key area: reporter protections and incentives.

Figure 6 shows the relationship of program goals and organizational culture to this key area.

TLR Reporting History and Biosafety Specialists' Views of Lab Culture Suggest Strong Confidentiality Protections, Data Deidentification, and Other Reporting Incentives Are Needed to Foster Trust in Reporting

Voluntary reporting to an SRS—especially of incidents that do not result in LAIs—would be a new expectation for some lab workers. As mentioned earlier, even the perception of a confidentiality breach can quash reporting.

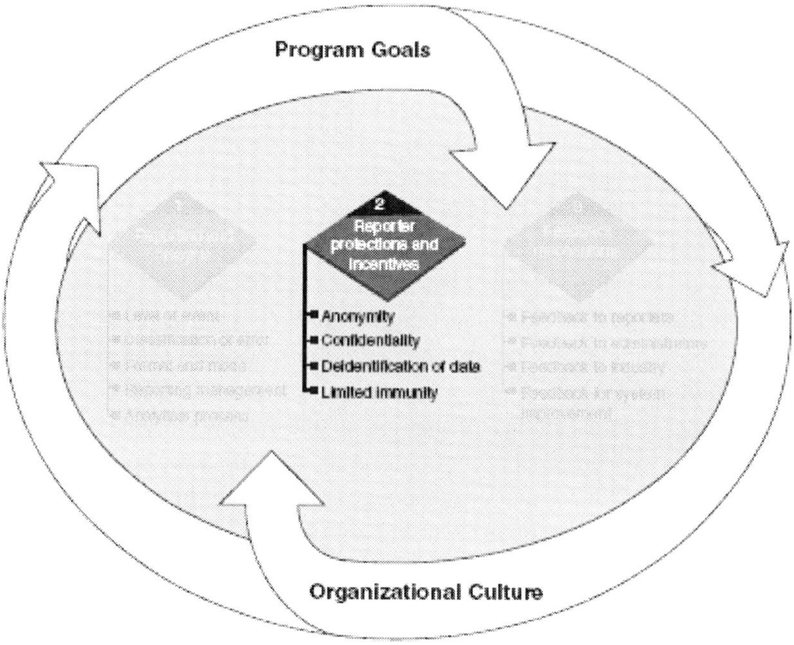

Source: GAO analysis of SRS evaluation literature.

Figure 6. Second Key Area—Reporter Protections and Incentives.

And given that entity information from the TLR reporting system was leaked to the press,[87] lab workers might have reason for concern about reporting similar incidents to a voluntary system. In addition, the literature and biosafety specialists noted, confidentiality concerns are among the barriers SRS managers will face in implementing a successful reporting program. Therefore, concerns about confidentiality suggest that a biological lab SRS will require strong confidentiality protections, data deidentification processes, and other incentives to encourage reporting, according to lessons learned. In addition, while the literature suggests anonymous reporting as one solution for minimizing confidentiality concerns, it is not an ideal one here. The complexity of biosafety issues would require a mechanism for follow-up with the worker or reporting entity because interpretation of the incident from a written report can often differ from interpretation of the incident from talking with the reporter, according to biosafety specialists.

Biosafety specialists also noted that developing trust in reporting has the potential to be problematic because of labs' existing reporting culture. For example, specialists noted the following influences on lab workers' likelihood of reporting accidents or incidents:

- realization that there is risk associated with laboratory work;
- difficulty recognizing that an incident has occurred, and knowing that this incident is reportable;
- disincentives for reporting, such as the threat of punishment for reporting or concerns about (1) the reputation of both the worker and the institution, (2) the potential loss of research funds, and (3) the fact that reporting may take time away from work; and
- lack of perceived incentives for reporting, such as the failure to see the value of reporting accidents or incidents, as well as the fact that lab work may be done alone, which does not provide an incentive for self-reporting of errors.

Given the confidentiality concerns and other difficulties of introducing a voluntary reporting system into the biological lab community, deidentification of safety reports takes on more importance. For example, according to biosafety specialists at one university, a primary concern with the establishment of their SRS was anonymity, especially for those in the agricultural labs. These researchers were concerned that if their identities became known, they could suffer from retaliation from organizations opposed to their research.

While the SRS managers chose to make the reports available to the public via the Web, they also deidentified the reports to prevent individuals outside the lab community from being able to identify individuals or specific labs. However, because the university research community is a small one and lab work is fairly specific, it is not overly difficult for those in the lab community to determine who was involved in an incident if a report mentions a particular pathogen and what was being done with it.

As a result, deidentification measures may have to go beyond simply removing reporter information. In addition, if deidentification measures are insufficient for maintaining confidentiality, workers and entities may need added incentives to encourage reporting in light of the fact that their identities may become known.

There are several incentives for the lab community to report, according to biosafety specialists. For example, deidentified SRS data can enhance the evidentiary foundation for biosafety research since it provides an extensive, heretofore unavailable data source. Such analyses benefit the overall lab community by providing greater evidentiary basis for risk based decisions for—or against—expensive or burdensome lab safety protocols. In addition, workers' trust in reporting can be developed over time at the local level, through rewarding, nonpunitive reporting experiences. The relationship workers have with the lab's safety staff is central to this effort, according to biosafety specialists.

Trust in an institution's Occupational Health Service, biosafety officer, or other official responsible for safety encourages workers to overcome ignorance, reluctance, or indifference to reporting. Biosafety specialists at one university credit the success of their nonpunitive SRS to the safety-focused relationship among the biosafety officer and lab staff. At first, according to these biosafety specialists, the researchers were afraid that SRS reports would be used to punish them academically or professionally. Over time, however, they saw the implementation of a nonpunitive system that had positive outcomes for safety improvements in the lab.

While biosafety specialists believed that development of a reporting culture might be difficult, they offered a number of suggestions for overcoming reporting barriers, including (1) developing a safety office in conjunction with the research staff, (2) ensuring continued interaction and shared conferences on safety issues with researchers and the biosafety office to show the value of reported information, and (3) reinforcing the importance of reporting by showing a concern for the individual that is exposed rather than focusing on punishment.

In addition, the CDC noted the importance of biosafety training, which is an important part of laboratory safety culture that has an impact on workers' ability to recognize and report safety issues. This type of continued support for reporting—as evidenced through positive feedback, awards, and nonpunitive experiences and training—fosters trust and willingness to report, according to lessons learned.

In the Context of Existing Information, Lessons Suggest Several Features for Feedback Mechanisms

Existing information about the potential goals for a biological lab SRS and the organizational culture of these labs suggest certain design and implementation features in the third key area: feedback mechanisms. Figure 7 shows the relationship of program goals and organizational culture to this key area.

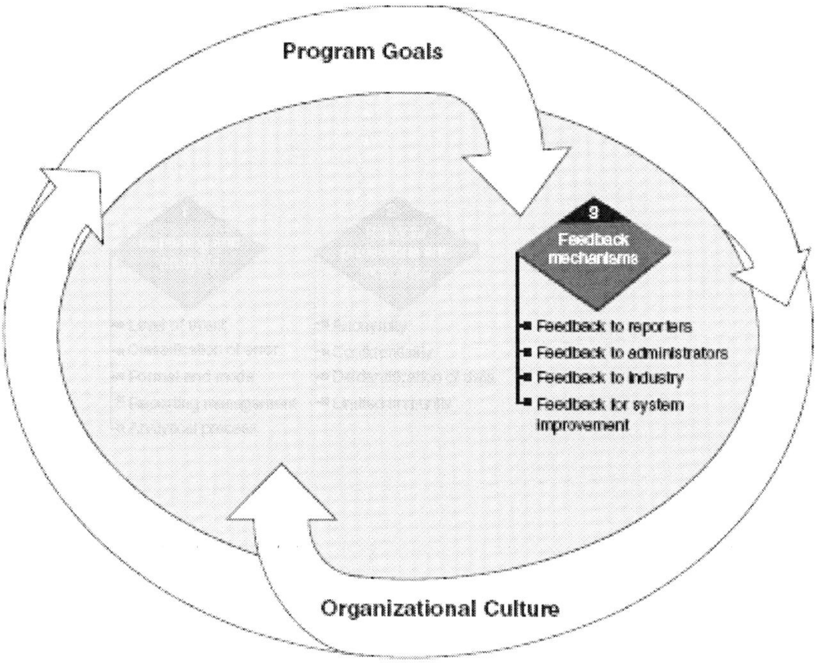

Source: GAO analysis of SRS evaluation literature.

Figure 7. Third Key Area—Feedback Mechanisms.

Lessons Suggest Industry-Level Entities, Such as the CDC or NIH, Can Facilitate Dissemination of SRS-Based Safety Information across the Lab Community

The CDC and NIH—as recognized authorities on working safely with infectious diseases—disseminate safety information to the entire lab community. For example, documents such as the BMBL and recombinant DNA guidelines provide the foundational principles for lab safety practices; they are updated periodically to reflect new information about infectious agents and routes of exposure. In addition, the CDC's MMWR reports provide alerts as emerging safety issues are identified. Lessons suggest that entities with industrywide visibility and recognized authority are ideally situated to ensure SRS data and safety improvement initiatives are disseminated across the industry. Such entities would be better positioned than individual labs, facilities, states, or others to disseminate SRS-based alerts or other safety reports in a way that reaches all labs. In addition, in order to counter the potential conflicts of interest that can arise with sharing data across labs, biosafety specialists we spoke with supported the notion of an "industry-level" entity for disseminating safety data. In particular, the specialists noted that the typical reporting relationship between the biosafety officer and lab management is not independent; this relationship might therefore inhibit sharing of safety data beyond the individual lab. Thus, a central, industry-level unit— responsible for collecting and disseminating SRS reports from either workers or organizations—minimizes such concerns and facilitates industrywide sharing of SRS data, according to lessons learned.

SRS data can also support training, which is a key component of biosafety. These data can provide the experiential basis for specific safety precautions. For example, one biosafety specialist noted that staff want to know this information in order to accept the need for precautions and procedures. Currently, there is no such experiential database; however, an industry-level entity could facilitate the creation and maintenance of such a database from SRS data.

Biosafety Specialists Note the Importance of Monitoring Safety Culture

Some of the biosafety specialists we spoke with noted the importance of ongoing monitoring of safety culture, for example, through a lab director's personal investment of time and direct observation and communication with lab workers. Without such observation and communication, as well as feedback from workers, managers will remain unaware of areas where the safety culture is likely to lead to serious problems. While specialists did not specifically note the need for formal evaluation to solicit this feedback, lessons learned suggest that evaluation is useful in this regard. Specifically, evaluation can help identify (1) problem areas in the safety culture and (2) where targeted outreach and training or program modification might lead to better reporting and safety improvement. Such evaluation is important in ensuring the system is working as effectively as possible, according to lessons from the literature and case studies.

CONCLUSION

Safety reporting systems (SRS) can be key tools for safety improvement efforts. Such systems increase the amount of information available for identifying systemic safety issues by offering a means through which workers can report a variety of events that shed light on underlying factors in the work environment that can lead to accidents. Our extensive review of SRS evaluation literature and case studies of SRS use in three industries provides an empirical, experience-based foundation for developing a framework for SRS design and implementation. This framework can be applied across a wide variety of industrial, organizational, professional, and cultural contexts. The industries we studied, despite their differences, shared similar experiences designing and using SRSs for safety improvement. The commonalities they shared provide the basis for our lessons—the pros and cons and successes and failures—relating to particular design and implementation choices across a wide variety of work environments. However, it is important to recognize the uniqueness of any work environment. The biological lab community is undoubtedly a unique working environment and blindly applying an SRS from one industry to the lab community would be a mistake. This observation underlies the leading finding among our lessons: in choosing the system features most appropriate for the environment in which the SRS will operate, consideration of program goals and organizational culture is essential. Such consideration provides the context for choosing features in three key areas of system design and implementation—reporting and analysis, reporter protections and incentives, and feedback mechanisms.

The Centers for Disease Control and Prevention (CDC) and Animal and Plant Health Inspection Service (APHIS) Select Agent Program (SAP) manage a mandatory reporting system for theft, loss, and release (TLR) of select agents. Although this system is compliance-based, it can be used— like the SRSs in our study—to identify systemic safety issues. In fact, the agencies have taken steps to use the system in this way. For example, the agencies have dedicated expert resources to manage the system, developed guidance to clarify reportable events and procedures to ensure reporter confidentiality, and used information from the system to provide feedback about safety issues to the select agent lab community. However, lessons from the literature and case studies suggest additional actions in assessment and the three key areas that could further improve reporting and the usefulness of the system as a source for safety data. These elements include an assessment of organizational culture, a lower threshold for reportable events, limited immunity provisions, and mechanisms for international lesson sharing and evaluation. Through these actions, efforts to identify areas for system improvement, target outreach and training, and encourage reporting could be supported.

While other industries have developed industrywide SRSs, one does not exist for the broader laboratory community. However, recognizing the potential of such a system for the laboratory community, an interagency task force on biosafety recommended it and Congress proposed legislation to develop one. While current safety guidance for biological labs is based on many years of experience working with infectious organisms and analyses of laboratory-acquired infections (LAI), there are some limitations to these data. For example, a widely recognized limitation is the high rate of under-reporting of LAIs. In addition, accident and illness data are incomplete, and reported information usually does not fully describe factors contributing to the LAIs. Such issues limit the amount of information available for identification of systemic factors that can lead to accidents. A national laboratorywide voluntary SRS that is accessible to all labs and designed around specific goals and organizational culture would facilitate collection of such data to inform safety improvements. Analysis of these data could support evidence-based modifications to lab practices and procedures, reveal problems with equipment use or design, and identify training needs and requirements.

Establishing such an SRS for the lab community, however, would require addressing some unique issues. Although our findings suggest that reporting systems should be tied to program goals and a clear sense of the organizational culture, this is problematic for biological labs because they are not a clearly identified or defined population. In addition, there is no agency or entity with the authority to direct such assessments across the entire lab community. Proposed federal legislation, if enacted, would establish a role for an SRS for the lab community to be administered by the Department of Health and Human Services (HHS) and the Department of Agriculture (USDA). If HHS and USDA are directed to develop such an SRS, certain features for the three key areas are suggested by existing studies, the CDC's and APHIS's experiences with the TLR reporting system, and biosafety specialists' knowledge of organizational culture in labs and experiences with safety reporting. Lessons developed from experiences with the National Institutes of Health's (NIH) prototype reporting system for its intramural research labs might inform design and implementation considerations as well. In addition, stakeholder involvement in goal setting is particularly important given the issues related to visibility and oversight of the broader lab population. The greater the stakeholder involvement, the greater the likelihood the perspectives of labs with varying environments and cultures will be represented. Stakeholders may also have knowledge of, and access to, labs that can support cultural assessments and encourage reporting. Such assessments are important for understanding differences in organizational cultures across the diverse types and levels of labs that could affect choices for system scope and features.

Until a cultural assessment is conducted, existing information about likely system goals and labs' organizational culture suggests certain features in the three key areas—reporting and analysis, reporter protections and incentives, and feedback mechanisms. With respect to reporting and analysis, a variety of factors suggest voluntary reporting for labs outside the Select Agent Program, including likely system goals for learning rather than enforcement and the need to collect information on incidents and hazards as opposed to serious accidents. In addition, the lab community's limited experience with this type of reporting, the diversity of lab environments, and uncertainty about the reporting population suggest an initially open classification scheme that allows workers to report events in their own words, using multimode (Web or postal) and open-format reporting options that are available to all workers. These options can facilitate reporting in such situations. Lastly, the advantages and disadvantages inherent in SRS administration at either the local or higher level suggest that dual reporting options may be necessary. Such options— present in different forms in all three case industries—allow workers to submit reports to whichever level is most comfortable for them. For example, workers would have the choice of whether to report to an internal, lab-managed reporting program that feeds data to a central authority or to an independent, externally managed SRS. Both of these reporting options will also require strong confidentiality protections, data deidentification, and other reporting incentives to foster trust in reporting. Finally, feedback mechanisms for disseminating safety data or recommendations and evaluations are needed to promote worker buy-in for reporting, identify areas for targeted outreach and training, and identify areas for system improvement.

Matters for Congressional Consideration

In developing legislation for a national reporting system for the biological laboratory community, Congress should consider provisions for the agency it designates as responsible for the system to take into account the following in design and implementation:

- include stakeholders in setting system goals;
- assess labs' organizational culture to guide design and implementation decisions;
- make reporting voluntary, with open-reporting formats that allow workers to report events in their own words and that can be submitted by all workers in a variety of modes (Web or postal), with the option to report to either an internal or external entity;

- incorporate strong reporter protections, data deidentification measures, and other incentives for reporting;
- develop feedback mechanisms and an industry-level entity for disseminating safety data and safety recommendations across the lab community; and
- ensure ongoing monitoring and evaluation of the safety reporting system and safety culture.

Recommendations for Executive Action

To improve the system for reporting the theft, loss, and release of select agents, we recommend that the Centers for Disease Control and Prevention and Animal and Plant Health Inspection Service Select Agent Program, in coordination with other relevant agencies, consider the following changes to their system:

- lower the threshold of event reporting to maximize collection of information that can help identify systemic safety issues,
- offer limited immunity protections to encourage reporting, and
- develop (1) mechanisms for sharing safety data for international lab safety improvement efforts and (2) processes for identifying reporting gaps and system evaluation to support targeted outreach and system modification.

Agency Comments and Our Evaluation

We provided a draft of this report to the Department of Transportation (DOT), HHS, INPO, NASA, NRC, USDA, and VA for review and comment. In written comments, the DOT, INPO, NASA, NRC, and VA agreed with our findings and conclusions and provided technical comments, which we addressed, as appropriate. The DOT's FAA and NASA also provided positive comments on the quality of our review. In particular, the FAA reviewer indicated that it was an excellent report that addressed the factors that should be considered by an organization planning to implement a safety reporting system. Similarly, the NASA reviewer noted that this was an excellent document describing the many aspects of safety reporting systems, and that it had captured the complexity and dynamic nature of the SRS approach to obtaining safety information from the frontline.

In written comments, the HHS noted that GAO's thorough case studies of long-standing industrywide safety reporting systems would be helpful when considering the important issue of reporting systems in biological laboratories. However, the HHS disagreed with two of our recommendations, and partially agreed with a third, to improve the theft, loss, and release (TLR) reporting system for select agents. Specifically, the HHS disagreed with our first recommendation—to lower the threshold for reportable events to maximize information collection—noting that their current mandatory reporting thresholds for the Select Agent Program (SAP) provides sufficiently robust information. While we appreciate the CDC and APHIS Select Agent Program's efforts to clarify reporting requirements to ensure all thefts, losses, and releases are reported, lowering reporting thresholds could further ensure all relevant reports are received. With lower reporting thresholds, questionable events are less likely to go unreported because of confusion about whether to report. Furthermore, we note that reporting below the currently established threshold could be voluntary, thereby offering registered entities a convenient, optional mechanism for sharing identified hazards. This is similar to the agencies' recently initiated, anonymous fraud, waste, and abuse reporting system. However, reporting to the TLR system would enable follow-up and feedback with the reporting lab because of its confidential, as opposed to anonymous, nature. Lastly, biosafety specialists we spoke with, as well as HHS staff involved in updating the BMBL, specifically noted the lack of available data for developing evidence-based biosafety guidelines. Data collected through the TLR system—especially if it is more comprehensive—could provide such data.

The HHS also disagreed with our second recommendation--to offer limited immunity protections to encourage reporting. While the HHS agrees that identification of safety issues is important, they believe they do not have statutory authority to offer limited immunity. The Public Health Security and Bioterrorism Preparedness and Response Act of 2002 required the Secretary of HHS to promulgate regulations requiring individuals and entities to notify HHS and others in the event of the theft, loss, or release of select agents and toxins. Violations of the Select Agent Regulations may result in criminal or civil money penalties. While we do not want to suggest that the HHS waive these penalties under a limited immunity provision, the Act sets maximum civil money penalties for Select Agent Regulations violations at $250,000 for individuals and $500,000 for entities, which provides the HHS Secretary, now delegated to the HHS Inspector General, discretion to charge penalties up to those maximum amounts. In addition, while reporting is required by law, individuals or entities may be concerned that reporting thefts, losses, or releases may lead to increased inspections by the CDC or referral to the Inspector General of the Department of Health and Human Services for investigation and possible penalties. Therefore, we recommend the CDC, in conjunction with other pertinent oversight agencies, examine whether adding limited immunity protections into the TLR reporting system would ease individuals' and entities' fears of reporting and encourage them to provide more complete information on thefts, losses, and releases. One possible way to incorporate limited immunity protections into the TLR reporting system would be to lower the civil money penalty for those individuals or entities who properly filed a TLR report should penalties be appropriate for the theft, loss, or release being reported. We believe the Secretary of HHS has sufficiently broad authority under the Public Health Security and Bioterrorism Preparedness and Response Act of 2002 to provide such immunity protections. The literature and our case studies identified limited immunity as a key incentive for reporting, and HHS' Trans-Federal Task Force on optimizing biosafety and biocontainment oversight noted the potential of the Aviation Safety Reporting System--and its associated immunity provisions--as a model for a national SRS for biological labs.

Lastly, the HHS partially agreed with the third recommendation. While the agency agreed with the recommendation to develop processes for identifying reporting gaps and system evaluation to support targeted outreach and system modification, they disagreed with the recommendation to share TLR data for international lab safety improvement efforts. In particular, the HHS notes its lack of authority to regulate foreign laboratories and suggests such activities might be better placed elsewhere in the CDC. As the literature and case studies illustrate, it is important to share safety lessons as broadly as possible. Sharing TLR lessons does not involve regulation of foreign labs, so additional authority is not required. Furthermore, the recommendation is directed to the CDC SAP because they manage the TLR system. If the CDC SAP wished to delegate the responsibility for sharing TLR lessons with the international lab community to another HHS entity, it would satisfy the intent of the recommendation.

The HHS also commented on the matters for congressional consideration, for example, suggesting additional matters that fall outside the scope of this review. The agency disagreed with GAO on several issues, such as (1) the scope of the recommendations, (2) the extent to which the biological lab industry might benefit from an SRS, (3) particular SRS features noted in the matters for congressional consideration, and (4) reporting thresholds and system management. The HHS also provided technical comments which we addressed, as appropriate.

In written comments, the USDA concurred with our recommendations, although they noted several disagreements in their detailed responses. With respect to our first recommendation—to lower reporting thresholds—the USDA noted, like the HHS, that (1) they believe the current reporting thresholds (providing 130 reports a year) are sufficiently robust and (2) APHIS's other monitoring and surveillance activities are sufficient for monitoring safety and security conditions in select agent questionable events are less likely to go unreported because of confusion about whether to report. Furthermore, we note that reporting below the currently established threshold could be voluntary, thereby offering registered entities a mechanism for sharing identified hazards in a system that would enable follow-up and feedback with reporters. Lastly, data collected through the TLR system—especially if it is more comprehensive—could provide data for updates to biosafety guidelines.

In response to our second recommendation—to offer limited immunity protections—the USDA, like the HHS, believes it lacks statutory authority to offer such protections. As noted above, we believe the Secretary of USDA has sufficiently broad authority under the Agricultural Bioterrorism Protection Act of 2002 to provide such immunity protections for the TLR reporting system. However, in recognition that such provisions might require coordination with other agencies, we added this clarification to the recommendations.

Lastly, in response to our third recommendation—to (1) share TLR data for international lab safety improvement efforts and (2) identify reporting gaps and conduct system evaluation—the USDA noted that they did not believe additional regulatory oversight was needed and that targeted education and safety training in high-risk areas would likely be more cost effective. Our recommendation does not suggest any additional regulatory oversight. It is focused on broadly sharing lessons learned from the TLR system and on identifying areas—through analysis of TLR data and evaluation—for targeted outreach and training and system modification. These actions are methods through which the USDA can better identify the "high-risk areas" the agency notes should be targeted for education and training. The USDA also noted that an example we provided of unreported LAIs demonstrates that these types of infections are infrequent. However, this is just one example of LAI underreporting and their consequences. As noted in the footnote prior to this example, in a review of LAI literature, the authors identified 663 cases of subclinical infections and 1,267 overt infections with 22 deaths. The authors also note that these numbers "represent a substantial underestimation of the extent of LAIs."[88] SRSs are key tools for bringing forward such safety information—currently recognized as substantially underreported—in order to benefit the entire industry.

Thomas J. McCool
Director, Applied Research and Methods

APPENDIX I: OBJECTIVES, SCOPE, AND METHODS

This appendix details the methods we used to identify lessons for designing and implementing an effective safety reporting system (SRS) from (1) the literature and (2) case studies of SRSs in the airline, commercial nuclear power, and health care industries; and apply those lessons to (3) assess the theft, loss, and release (TLR) reporting system for the Select Agent Program and (4) suggest design and implementation considerations for a national SRS for all biological labs.

To develop lessons from the literature, we used an iterative approach to search several venues (academic journals, agency and organization publications, and grey literature) for literature related to human factors, safety science, and SRS evaluation. We reviewed the publications generated through automated searches to identify (1) search terms for additional automated queries and (2) citations for publications that might be within our scope of interest.

We ended the formal search for additional literature after reaching saturation in the publications generated from our search (i.e., no or few new publications).

The literature we reviewed generally fell into one of two categories—safety science (including human factors and organizational safety) literature and descriptions of SRS features and evaluations. The safety science literature serves as background information and was also used to develop familiarity with safety science terms and theories required for our assessment of the SRS evaluation literature. The literature related to SRS features and evaluations was used to develop lessons for the first objective. We assessed the SRS evaluation literature for both methodological rigor and findings related to SRS design and implementation.

For the methodological review, we assessed the appropriateness of the methods relative to the study objectives for all articles, and a sample (about half) received a secondary, independent review of methodological rigor. Studies that met our standards of methodological rigor were incorporated into the assessment, and findings related to system goals, cultural considerations, reporting and analysis features, reporter protections and incentives, and feedback mechanisms were coded to identify effective features and processes for SRS design and implementation. See the Bibliography of Articles Used to Develop SRS Lessons from the Literature for a list of the literature used to develop these lessons.

To develop lessons from case studies of three industries, we (1) reviewed studies and documentation on a variety of SRSs in the three industries; (2) interviewed agency and organization officials knowledgeable about safety science and human factors engineering, reporting systems, and their own SRS programs; and (3) attended a variety of SRS and safety conferences. We chose to focus on the aviation, commercial nuclear power, and health care industries because they are moderate- to high-risk industries that represent a variety of (1) organizational cultures, (2) length of experience using SRSs for safety improvement, and (3) feature and design choices in their SRS programs. While we collected information on a wide variety of safety reporting programs and systems in these industries—and in some cases comment on these different programs—we primarily developed our lessons from one reporting program in each of the three industries. Specifically, we developed lessons from the Federal Aviation Administration's (FAA) National Aeronautic and Space Administration (NASA)-run Aviation Safety Reporting System (ASRS) in aviation, the Institute of Nuclear Power Operation's (INPO®) Significant Event Evaluation-Information Network (SEE-IN®) system in commercial nuclear power, and the VA's internally managed Patient Safety Information System (PSIS) and NASA-managed Patient Safety Reporting System (PSRS) in VA health care. We chose to focus on these systems because they represent fairly long-standing, nonregulatory, domestic, industrywide or servicewide reporting programs.

For example, NASA's ASRS has been in operation for 34 years; INPO's SEE-IN, for 30 years; and VA's PSIS and PSRS, for 10 years. Although we primarily developed our lessons from these key SRSs, we also collected information on other notable SRSs in the industries, including the Nuclear Regulatory Commission's (NRC) Allegations Program, the FAA's Aviation Safety Action Program (ASAP), and the Agency for Healthcare Research and

Quality's (AHRQ) Patient Safety Organizations (PSO) program, among others.

To assess the TLR reporting system, we interviewed agency officials, reviewed agency and other documentation, and applied lessons from the literature and case studies to these findings.

Specifically, using a standard question set, we interviewed HHS officials from the Coordinating Center for Infectious Disease, Office of Health and Safety, and Division of Select Agents and Toxins, and received responses to our question set from the USDA's Animal and Plant Health Inspection Service (APHIS).

In addition, we attended an agency conference on select agent reporting and reviewed documents from this conference and from the National Select Agent Registry (NSAR) Web site, detailing TLR reporting requirements and scenarios.

We also reviewed GAO testimony and reports on previously identified TLR reporting issues. Using the lessons for SRS design and implementation derived from the literature and case studies, we applied these criteria to identify areas for TLR improvements.

To propose design and implementation considerations for a national biological laboratory reporting system, we reviewed studies and other reports on biosafety, interviewed HHS officials and domestic and international biosafety specialists, attended conferences on biosafety and incident reporting, and applied lessons from the literature and case studies to these findings. We interviewed HHS officials and biosafety specialists to get a sense of the culture-related context for, and potential barriers to, an SRS for biological labs.

Specifically, we used a standardized question set to gather specialists' views about overall design and implementation considerations for a labwide reporting program, as well as how lab culture and safety orientation (1) vary by level and type of lab; (2) affect reporting under current requirements; and (3) might affect reporting to a national biological lab SRS.

We conducted this performance audit from March 2008 through September 2010 in accordance with generally accepted government auditing standards.

Those standards require that we plan and perform the audit to obtain sufficient, appropriate evidence to provide a reasonable basis for our findings and conclusions based on our audit objectives. We believe that the evidence obtained provides a reasonable basis for our findings and conclusions based on our audit objectives.

APPENDIX II: SUMMARY OF LESSONS FROM THE LITERATURE AND CASE STUDIES

Area	Lessons from the literature	Lessons from case studies
System goals and organizational culture	(1) Define overarching program goals and subgoals up front. (2) Involve stakeholders (e.g., management, industry groups, associations, and workers) in development of program goals and SRS design to increase support among key populations. (3) Assess organizational culture to guide system design choices in the three key areas. (4) Ensure that reporters and system administrators receive adequate training regarding the function and application of the reporting system.	(1) Assessment, dedicated resources, and management focus are needed to understand and improve safety culture. (1a) Assessing safety culture can alert management to workplace safety issues. (1b) Improving safety culture requires dedicated resources, including time, training, and staff investment. (1c) Changing safety culture requires management focus.
Reporting and analysis	Level of event (1) Base the decision for mandatory or voluntary reporting on (a) the level of event of interest and (b) whether the SRS will be used primarily for enforcement or learning. (2) Set reporting thresholds that are not so high that reporting is curtailed, nor so low that the system is overwhelmed by the number and variety of reportable events.	(2) Broad reporting thresholds, experience-driven classification schemes, and processing at the local level can be useful SRS features in industries new to safety reporting. (2a) Broad thresholds and open reporting are useful features when starting an SRS.
	Event classification (1) Develop classification schemes and associated terms that are clear, easy to understand, and easy to use by drawing on terms already well understood in the industry. (2) Test whether classification terms are clearly understood by different groups in the organization. (3) Allow sufficient flexibility to (a) avoid narrowing the scope of reporting in a way that limits all events of	(2b) Encouraging workers to report incidents in their own words facilitates reporting initially.

Appendix II: (Continued)

Area	Lessons from the literature	Lessons from case studies
	interest at the chosen level of event, (b) allow different sites—if multiple sites will be reporting to the same system—to adapt fields and elements to match their own organizational culture, and (c) capture different types of events and precursors as they can change over time. (4) Develop a classification scheme that best suits analytical requirements and the comfort level of the organizational culture with safety reporting and safety event terms.	
	Mode and format (1) Base decisions about report mode on (a) the accessibility of the mode to the reporting population and (b) workers' concerns about and willingness to report. (2) Base decisions about report formats on the (a) type of data needed for analysis, (b) capabilities of the reporting population, and (c) maturity of existing safety event classification schemes within the industry.	
	System administration (1) Base the decision for internal or external system administration on (a) workers' degree of concern over punishment and confidentiality and (b) the availability of internal expertise and resources to analyze and encourage reporting. (2) Base decisions about who will be allowed to report on (a) awareness of reporting hierarchies and (b) the type of information desired for analysis.	(2c) Reporting options with some local-levelprocessing facilitates reporting initially.
	Analysis (1) Use a report prioritization process to quickly and efficiently address key safety issues as they arise.	

Area	Lessons from the literature	Lessons from case studies
	(2) Align analysis decisions with (a) report formats, (b) system administration and location of technical expertise, and (c) availability of other relevant data needed for analysis.	
Reporter protections and incentives	Confidentiality and anonymity (1) Base the choice between anonymity and confidentiality on (a) organizational culture, especially workers' degree of concern about punishment and confidentiality, and (b) the amount of detail required for analysis and whether it can be collected without follow-up. (2) Consider a hybrid system in which confidential and anonymous reporting are used simultaneously if there is conflict between organizational culture and data need. Data deidentification (1) Develop data deidentification measures to support confidentiality and data-sharing efforts. Limited immunity (1) Consider limited immunity provisions to increase the reporting incentive.	(3) Strong legal protections and incentives encourage reporting and help prevent confidentiality breaches.
Feedback mechanisms	Feedback (1) Provide direct feedback to reporters to foster worker-specific buy-in for reporting. (2) Provide regular, timely, and routine feedback—for example in the form of newsletters, e-mail alerts, Web sites, and searchable databases—to support overall organizational buy-in for reporting. (3) Provide positive feedback to managers who receive a high volume of reports to demonstrate the importance of reporting and counteract the perception that error reporting reflects poorly on management. Evaluation	(4) A central, industry-level unit facilitates lesson sharing and evaluation.

Appendix II: (Continued)

Area	Lessons from the literature	Lessons from case studies
	(1) Use the data to identify reporting gaps for targeted outreach and training. (2) Evaluate the effectiveness of the SRS to support ongoing modification and improvement.	

Source: GAO analysis of SRS literature and case studies.

BIBLIOGRAPHY OF ARTICLES USED TO DEVELOP SRS LESSONS FROM THE LITERATURE

Aagaard, L., B. Soendergaard, E. Andersen, J. P. Kampmann and E. H. Hansen. "Creating Knowledge About Adverse Drug Reactions: A Critical Analyis of the Danish Reporting System from 1968 to 2005." *Social Science & Medicine*, vol. 65, no. 6 (2007): 1296-1309.

Akins, R. B. "A Process-centered Tool for Evaluating Patient Safety Performance and Guiding Strategic Improvement." In *Advances in Patient Safety: From Research to Implementation*, 4,109-125. Rockville, Md: Agency for Healthcare Research and Quality, 2005.

Anderson, D. J. and C. S. Webster. "A Systems Approach to the Reduction of Medication Error on the Hospital Ward." *Journal of Advanced Nursing*, vol. 35, no. 1 (2001): 34-41.

Arroyo, D. A. "A Nonpunitive, Computerized System for Improved Reporting of Medical Occurrences." In *Advances in Patient Safety: From Research to Implementation*, 4, 71-80. Rockville, Md.: Agency for Healthcare Research and Quality, 2005.

Bakker, B. "Confidential Incident Reporting Systems For Small Aviation Communities on a Voluntary Basis." *Aviation Safety* (1997): 790-720.

Baldwin, I., U. Beckman, L. Shaw and A. Morrison. "Australian Incidence Monitoring Study in Intensive Care: Local Unit Review Meetings and Report Management." *Anaesthesia and Intensive Care*, vol. 26, no. 3 (1998): 294-297.

Barach, P. and S. D. Small. "Reporting and Preventing Medical Mishaps: Lessons from Non-medical Near Miss Reporting Systems." *British Medical Journal*, 320 (2000): 759-763.

Battles, J. B., H. S. Kaplan, T. W. Van der Schaaf and C. E Shea. "The Attributes of Medical Event-Reporting Systems: Experience with a Prototype Medical Event-Reporting System for Transfusion Medicine." *Archives of Pathology & Laboratory Medicine*, vol. 122, no. 3 (1998): 231- 238.

Battles, J. B., N. M. Dixon, R. J. Borotkanics, B. Rabin-Fastmen and H. S. Kaplan. "Sensemaking of Patient Safety Risks and Hazards." *Health Services Research*, vol. 41, no. 4 (2006): 1555-1575.

Beaubien, J. M. and D. P. Baker. "A Review of Selected Aviation Human Factors Taxonomies, Accident/Incident Reporting Systems, and Data Reporting Tools." *International Journal of Applied Aviation Studies*, vol. 2, no. 2 (2002): 11-36.

Beckett, M. K., D. Fossum, C. S. Moreno, J. Galegher and R. S. Marken. "A Review of Current State-Level Adverse Medical Event Reporting Practices Toward National Standards." *RAND Health: Technical Report.* 2006.

Berkowitz, E. G., M. E. Ferrant, L. B. Goben, K. E. McKenna, and J. L. Robey. "Evaluation of Online Incident Reporting Systems." *Duke University School of Nursing* (2005): 1-27.

Billings, C. E. Some "Hopes and Concerns Regarding Medical Event-Reporting Systems." *Archives of Pathology & Laboratory Medicine*, vol. 122, no. 3 (1998): 214-215.

Bloedorn, E. *Mining Aviation Safety Data: A Hybrid Approach*. The MITRE Corporation, 2000.

Braithwaite, J., M. Westbrook, and J. Travaglia. "Attitudes Toward the Large-scale Implementation of an Incident Reporting System." *International Journal for Quality in Health Care*, vol. 20, no. 3 (2008): 184-191.

Centers for Disease Control and Prevention. *CDC Workbook on Implementing a Needlestick Injury Reporting System*. 2008.

Chidester, T. R. *Voluntary Aviation Safety Information-Sharing Process: Preliminary Audit of Distributed FOQA and ASAP Archives Against Industry Statement of Requirements*. DOT/FAA/AM-07/7. A report prepared at the request of the Federal Aviation Administration. 2007.

Clarke, J. R. "How a System for Reporting Medical Errors Can and Cannot Improve Patient Safety." *The American Surgeon*, vol. 72, no. 11 (2006): 1088-1091.

Connell, L. J. *Cross-Industry Applications of a Confidential Reporting Model*, 139-146. Washington D.C.: National Academy of Engineering, 2004.

Council of Europe: Expert Group on Safe Medication Practices. *Creation of a Better Medication Safety Culture in Europe: Building Up Safe Medication Practices*. P-SP-PH/SAFE. 2006.

Dameron, J. and L. Ray. *Hospital Adverse Event Reporting Program: an Initial Evaluation.* Oregon Patient Safety Commission, 2007.

Daniels, C. and P. Marlow. *Literature Review on the Reporting of Workplace Injury Trends*. HSL/2005/36. Buxton, Derbyshire, UK: Health and Safety Laboratory, 2005.

Department of Defense. Assistant Secretary of Defense for Health Affairs. *Military Health System Clinical Quality Assurance Program Regulation.* DoD 6025.13-R. 2004.

Desikan, R., M. J. Krauss, W. Claiborne Dunagan, E. C. Rachmiel, T. Bailey, and V. J. Fraser. "Reporting of Adverse Drug Events: Examination of a Hospital Incident Reporting System." In *Advances in Patient Safety: From Research to Implementation*, 1, 145-160. Rockville, Md.: Agency for Healthcare Research and Quality, 2005.

Evans, S. M., J. G. Berry, B. J. Smith, A. Esterman, P. Selim, J. O'Shaughnessy and M. DeWit. "Attitudes and Barriers to Incident Reporting: A Collaborative Hospital Study." *Quality and Safety in Health Care*, 15 (2006): 39-43.

Fernald, D. H., W. D. Pace, D. M. Harris, D. R. West, D. S. Main and J. M. Westfall. "Event Reporting to a Primary Care Patient Safety Reporting System: A Report from the ASIPS Collaborative." *Annals of Family Medicine*, vol. 2, no. 4 (2004): 327-332.

Flack, M., T. Reed, J. Crowley, and S. Gardner. "Identifying, Understanding, and Communicating Medical Device Use Errors: Observations from an FDA Pilot Program." In *Advances in Patient Safety: From Research to Implementation*, 3, 223-233. Rockville, Md.: Agency for Healthcare Research and Quality, 2005.

Flink, E., C. L. Chevalier, A. Ruperto, P. Dameron, F. J. Heigel, R. Leslie, J. Mannion and R. J. Panzer. "Lessons Learned from the Evolution of Mandatory Adverse Event Reporting Systems." In *Advances in Patient Safety: From Research to Implementation*, 1-4, 135-151. Rockville, Md.: Agency for Healthcare Research and Quality, 2005.

Flowers, L. and T. Riley. *State-based Mandatory Reporting of Medical Errors: An Analysis of the Legal and Policy Issues.* National Academy for State Health Policy, 2001.

Frey, B., V. Buettiker, M. I. Hug, K. Waldvogel, P. Gessler, D. Ghelfi, C. Hodler and O. Baenziger. Does Critical Incident Reporting Contribute to

Medication Error Prevention?" *European Journal of Pediatrics*, vol. 161, no. 11 (2002): 594-599.

Ganter, J. H., C. D. Dean, and B. K. Cloer. *Fast Pragmatic Safety Decisions: Analysis of an Event Review Team of the Aviation Safety Action Partnership.* SAND2000-1134. Albuquerque, N.M.: Sandia National Laboratories, 2000.

Gayman, A. J., A. W. Schopper, F. C. Gentner, M. C. Neumeier, and W. J. Rankin. *Crew Resource Management (CRM) Anonymous Reporting System (ARS) Questionnaire Evaluation.* CSERIAC Report CSERIACRA-96-003. 1996.

Global Aviation Information Network (GAIN) Working Group E. "A Roadmap to a Just Culture: Enhancing the Safety Environment." *Flight Safety Digest*, vol. 24, no. 3 (2005): 1-48.

Grant, M. J. C. and G. Y. Larsen. "Effect of an Anonymous Reporting System on Near-miss and Harmful Medical Error Reporting in a Pediatric Intensive Care Unit." *Journal of Nursing Care Quality*, vol. 22, no. 3 (2007): 213-221.

Harper, M. L. and R. L. Helmreich. "Identifying Barriers to the Success of a Reporting System. Advances." In *Advances in Patient Safety: From Research to Implementation*, 3, 167-179. Rockville, Md.: Agency for Healthcare Research and Quality, 2004.

Hart, C. A. "Stuck on a Plateau: A Common Problem." In *Accident Precursor Analysis and Management: Reducing Technological Risk Through Diligence*, 147-154. Phimister, J. R., V. M. Bier, and H. C. Kunreuther, Eds. Washington, D.C.: National Academies Press, 2004

Holden, R. J. and B.-T. Karsh. "A Review of Medical Error Reporting System Design Considerations and a Proposed Cross-Level Systems Research Framework." *Human Factors; the Journal of the Human Factors Society*, vol. 49, no. 2 (2007): 257-276.

Holzmueller, C. G., P. J. Pronovost, F. Dickman, D. A. Thompson, A. W. Wu, L. H. Lubomski, M. Fahey, D. M. Steinwachs, L. Engineer, A. Jaffrey, et al. "Creating the Web-based Intensive Care Unit Safety Reporting System."*Journal of the American Medical Informatics Association*, vol. 12, no. 2 (2005): 130-139.

International Atomic Energy Agency. *The IAEA/NEA Incident Reporting System: Using Operational Experience to Improve Safety.*

International Atomic Energy Agency. *Safety Culture*. A Report by the International Nuclear Safety Advisory Group. Safety Series: 75-INSAG-4. International Nuclear Safety Advisory Group, 1991.

Johnson, C. "Software Tools to Support Incident Reporting in Safety-Critical Systems." *Safety Science*, vol. 40, no. 9 (2002): 765-780.

Kaplan, H. and P. Barach. "Incident Reporting: Science or Protoscience? Ten Years Later." *Quality and Safety in Health Care*, vol. 11, no. 2 (2002): 144-145.

Kaplan, H., J. Battles, Q. Mercer, M. Whiteside and J. Bradley. *A Medical Event Reporting System for Human Errors in Transfusion Medicine*, 809-814. Lafayette, Ind.: USA Publishing, 1996.

Kaplan, H.S. and B. R. Fastman. "Organization of Event Reporting Data for Sense Making and System Improvement." *Quality and Safety in Health Care*, vol. 12 (2003): ii68-ii72.

Khuri, S. F. "Safety, Quality, and the National Surgical Quality Improvement Program." *The American Surgeon*, vol. 72, no. 11 (2006): 994-998.

Krokos, K. J. and D. P. Baker. *Development of a Taxonomy of Causal Contributors for Use with ASAP Reporting Systems*, 1-59. American Institutes for Research, 2005.

Leape, L. L. Reporting of Adverse Events. *The New England Journal of Medicine*, vol. 347, no. 20 (2002): 1633-1639.

Lee, R. The Australian Bureau of Air Safety Investigation, in Aviation Psychology: A Science and a Profession, 229-242. U.K.: Ashgate Publishing, 1998.

Martin, S. K., J. M. Etchegaray, D. Simmons, W. T. Belt and K. Clark. Development and "Implementation of The University of Texas Close Call Reporting System." *Advances in Patient Safety*, vol. 2 (2005): 149-160.

Morters, K., and R. Ewing. "The Introduction of a Confidential Aviation Reporting System into a Small Country." *Human Factors Digest*, vol. 13 (1996): 198-203.

Murff, H. J., D. W. Byrne, P. A. Harris, D. J. France, C. Hedstrom, and R. S. Dittus. "'Near-Miss' Reporting System Development and Implications for Human Subjects Protection." In *Advances in Patient Safety: From Research to Implementation*, 3, 181-193. Rockville, Md.: Agency for Healthcare Research and Quality, 2005.

Nakajima,K., Y. Kurata and H. Takeda. "A Web-based Incident Reporting System and Multidisciplinary Collaborative Projects for Patient Safety in a Japanese Hospital." *Quality and Safety in Health Care*, vol. 14 (2005): 123-129.

National Academy of Engineering of the National Academy. 2004. "The Accident Precursors Project: Overview and Recommendations." In

Accident Precursor Analysis and Management: Reducing Technological Risk Through Diligence, 1-34. Phimister, J. R., V. M. Bier, and H. C. Kunreuther, Eds. Washington, D.C.: National Academies Press, 2004.

National Aeronautics and Space Administration. *ASRS: The Case for Confidential Incident Reporting Systems.* Pub 60. 2001.

National Transportation Safety Board. *Current Procedures for Collecting and Reporting U.S. General Aviation Accident and Activity Data.* NTSB/SR-05/02. 2005.

Nguyen, Q.-T., J. Weinberg, and L. H. Hilborne. "Physician Event Reporting: Training the Next Generation of Physicians." In *Advances in Patient Safety: From Research to Implementation*, 4, 353-360. Rockville, Md.: Agency for Healthcare Research and Quality, 2005

Nielsen, K. J., O. Carstensen, K. and Rasmussen. "The Prevention of Occupational Injuries in Two Industrial Plants Using an Incident Reporting Scheme." *Journal of Safety Research*, vol. 37 (2006): 479-486.

Nørbjerg, P. M. "The Creation of an Aviation Safety Reporting Culture in Danish Air Traffic Control." *CASI* (2003): 153-164.

Nosek, Jr., R. A., J. McMeekin, and G. W. Rake. 2005. "Standardizing Medication Error Event Reporting in the U.S. Department of Defense." In *Advances in Patient Safety: From Research to Implementation*, 4, 361-374. Rockville, Md.: Agency for Healthcare Research and Quality, 2005.

O'Leary, M. J and S. L. Chappell. *Early Warning: Development of Confidential Incident Reporting Systems.* NASA Center for AeroSpace Information, 1996.

Page, W. D., E. W. Staton, G. S. Higgins, D. S. Main, D. R. West and D. M. Harris. "Database Design to Ensure Anonymous Study of Medical Errors: A Report from the ASIPS Collaborative." *Journal of the American Medical Informatics Association*, vol. 10, no. 6 (2003): 531-540.

Patankar, M. S. and J. Ma. "A Review of the Current State of Aviation Safety Action Programs in Maintenance Organizations." *International Journal of Applied Aviation Studies*, vol. 6. no. 2 (2006): 219-233.

Phillips, R. L., S. M. Dovey, J. S. Hickner, D. Graham and M. Johnson. "The AAFP Patient Safety Reporting System: Development and Legal Issues Pertinent to Medical Error Tracking and Analysis." In *Advances in Patient Safety: From Research to Implementation,* 3, 121-134. Rockville, Md.: Agency for Healthcare Research and Quality, 2005.

Phimister, J. R., V. M. Bier and H. C. Kunreuther. "Flirting with Disaster." *Issues in Science and Technology* (2005).

Pronovost, P. J., B. Weast, C. G. Holzmueller, B. J. Rosenstein, R. P. Kidwell, K. B. Haller, E. R. Feroli, J. B. Sexton, and H. R. Rubin. "Evaluation of the Culture of Safety: Survey of Clinicians and Managers in an Academic Medical Center." *Quality and Safety in Health Care*, vol. 12, no. 6 (2003): 405-410.

Ramanujam, R., D. J. Keyser and C. A. Sirio. "Making a Case for Organizational Change in Patient Safety Initiatives." In *Advances in Patient Safety: From Research to Implementation*, 2, 455-465. Rockville, Md.: Agency for Healthcare Research and Quality, 2005.

Raymond, B. and R. M. Crane. *Design Considerations for Patient Safety Improvement Reporting System.* Kaiser Permanente Institute for Health Policy, NASA Aviation Safety Reporting System, and The National Quality Forum, 2000.

Rejman, Michael H. *Confidential Reporting Systems and Safety-Critical Information*, 397-401. Columbus, Ohio: Ohio State University, 1999.

Reynard, W.D., C.E. Billings, E.S. Cheaney and R. Hardy. *The Development of the NASA Aviation Safety Reporting System*, Pub 34. NASA Reference Publication, 1986.

Ricci, M., A. P. Goldman, M. R. de Leval, G. A. Cohen, F. Devaney and J. Carthey. "Pitfalls of Adverse Event Reporting in Pediatric Cardiac Intensive Care." *Archives of Disease in Childhood*, 89 (2004): 856-859.

Ruchlin, H. S., N. L. Dubbs, M. A. Callahan and M. J. Fosina. "The Role of Leadership in Installing a Culture of Safety: Lessons from the Literature." *Journal of Healthcare Management*, vol. 49, no. 1 (2004): 47-59.

Schleiffer, S. C. "We Need to Know What We Don't Know." *International Air Safety Seminar*, 35 (2005): 333-340.

Snijders, C., R. A. van Tingen, A .Molendijk, and W. P. F. Fetter. "Incidents and Errors in Neonatal Intensive Care: A Review of the Literature." *Archives of Disease in Childhood, Fetal and Neonatal Editon*, vol. 92, no. 5 (2007): 391-398.

Staender, S., J. Davies, B. Helmreich, B. Sexton and M. Kaufmann. "The Anaethesia Critical Incident Reporting System: An Experience Based Dataset." *International Journal of Medical Informatics*, vol. 47, no. 1-2 (1997): 87-90.

Stainsby, D., H. Jones, D. Asher, C. Atterbury, A. Boncinelli, L. Brant, C. E. Chapman, K. Davison, R. Gerrard, A. Gray et al. "Serious Hazards of Transfusion: A Decade of Hemovigilance in the UK." *Transfusion Medicine Reviews*, vol. 20, no. 4 (2006): 273-282.

Stalhandske, E., J. P. Bagian, and J. Gosbee. "Department of Veterans Affairs Patient Safety Program." *American Journal of Infection Control*, vol. 30, no. 5 (2002): 296-302.

Stump, L. S. "Re-engineering the Medication Error-Reporting Process: Removing the Blame and Improving the System." *American Journal of Health-System Pharmacy*, vol. 57, no. 24 (2000): S10-S17.

Suresh, G., J. D. Horbar, P. Plsek, J. Gray, W. H. Edwards, P. H. Shiono, R. Ursprung, J. Nickerson, J. F. Lucey, and D. Goldmann. "Voluntary Anonymous Reporting of Medical Errors for Neonatal Intensive Care." *Pediatrics*, 113 (2004): 1609-1618.

Tamuz, M. "Learning Disabilities for Regulators: The Perils of Organizational Learning in the Air Transportation Industry." *Administration & Society*, 33 (2001): 276-302.

Tamuz, M. and E. J. Thomas. "Classifying and Interpreting Threats to Patient Safety in Hospitals: Insights from Aviation." *Journal of Organizational Behavior*, 27 (2006): 919-940.

Taylor, J. A., D. Brownstein, E. J. Klein, and T. P. Strandjord. "Evaluation of an Anonymous System to Report Medical Errors in Pediatric Inpatients." *Journal of Hospital Medicine*, vol. 2, no. 4 (2007): 226-233.

Tuttle, D., R. Holloway, T. Baird, B. Sheehan, and W. K. Skelton. "Electronic Reporting to Improve Patient Safety." *Quality and Safety in Health Care*, 13 (2004): 281-286.

U.S. Department of Energy. Office of Environment, Safety and Health. *Occurrence Reporting and Processing of Operations Information.* DOE 231.1-2. 2003.

U.S. Nuclear Regulatory Commission. Working Group on Event Reporting. *Final Report of the Working Group on Event Reporting.* 2001.

Ulep, S. K. and S. L. Moran. 2005. "Ten Considerations for Easing the Transition to a Web-based Patient Safety Reporting System." *In Advances in Patient Safety: From Research to Implementation*, 3, 207-222. Rockville, Md.: Agency for Healthcare Research and Quality.

Underwood, P. K (LTC). *Medical Errors: An Error Reduction Initiative*,1- 75. U.S. Army—Baylor University Graduate Program in Healthcare Adminstration, 2001.

van der Schaaf, T. W. and L. Kanse. *Checking for Biases in Incident Reporting*, 119-126. Washington D.C.: National Academy of Engineering, 2004.

van der Schaaf, T. W. "Medical Applications of Industrial Safety Science." *Quality and Safety in Health Care*, vol. 11, no. 3 (2002): 205-206.

Wallace, B., A. Ross and J. B. Davies. "Applied Hermeneutics and Qualitative Safety Data: The CIRAS Project." *Human Relations*, vol. 56, no. 5 (2003): 587-607.

Webster, C. S. and D. J. Anderson. "A Practical Guide to the Implemenation of an Effective Incidence Reporting Scheme to Reduce Medication Error on the Hospital Ward." *International Journal of Nursing Practice*, vol. 8, no. 4 (2002): 176-183.

Weinberg, J., L. H. Hilborne, Q.-T. Nguyen. "Regulation of Health Policy: Patient Safety and the States." In *Advances in Patient Safety: From Research to Implementation*, 1, 405-422. Rockville, Md.: Agency for Healthcare Research and Quality, 2005.

Weiner, B. J., C. Hobgood, and M. Lewis. "The Meaning of Justice in Safety Incident Reporting." *Social Science & Medicine*, vol. 66, no. 2 (2008): 403- 413.

Wiegmann, D. A. and T.L. von Thaden. *The Critical Event Reporting Tool (CERT); Technical Report.* University of Illinois at Urbana-Champaign: Aviation Research Lab, Institute of Aviation. ARL-01-7/FAA-01-2. 2001.

Wilf-Miron, R., I. Lewenhoff, Z. Benyamini, and A. Aviram. "From Aviation to Medicine: Applying Concepts of Aviation Safety to Risk Management in Ambulatory Care." *Quality and Safety in Health Care*, vol.12, no. 1 (2003): 35-39.

Wu, A. W, P. Pronovos and L. Morlock. "ICU Incident Reporting Systems." *Journal of Critical Care*, vol. 17, no. 2 (2002): 86-94.

Yong, K. "An Independent Aviation Accident Investigation Organization in Asia Pacific Region—Aviation Safety Council of Taiwan." *International Air Safety Seminar Proceedings*, 173-180. 2000.

BIBLIOGRAPHY OF OTHER LITERATURE USED IN THE REPORT

Barhydt, R. and C. A. Adams. Human Factors Considerations for Area Navigation Departure and Arrival Procedures. A report prepared for NASA. 2006.

Besser, R. E. *Oversight of Select Agents by the Centers for Disease Control and Prevention.* Testimony before Subcommittee on Oversight and Investigations, Committee on Energy and Commerce, United States House of Representatives. 2007

Center for Biosecurity. University of Pittsburgh Medical Center. *Response to the European Commission's Green Paper on Bio-preparedness.* 2007.

Gronvall G. K., J. Fitzgerald, A. Chamberlain, T. V. Inglesby, and T. O'Toole. "High-Containment Biodefense Research Laboratories: Meeting Report and Center Recommendations." *Biosecurity and Bioterrorism: Biodefense Strategy, Practice, and Science*, vol. 5, no. 1 (2007): 75-85.

Gronvall G. K. *Germs, Viruses, and Secrets: The Silent Proliferation of Bio-Laboratories in the United States.* University of Pittsburgh Medical Center, Center for Biosecurity, 2007.

Gronvall G. K., J. Fitzgerald, T.V. Inglesby, and T. O'Toole. "Biosecurity: Responsible Stewardship of Bioscience in an Age of Catastrophic Terrorism." *Biosecurity and Bioterrorism: Biodefense Strategy, Practice, and Science*, vol. 1, no. 1 (2003): 27-35.

Hallbert, B., R. Boring, D. Gertman, D. Dudenhoeffer, A. Whaley, J. Marble, J. Joe, and E. Lois. *Human Event Repository and Analysis (HERA) System, Overview*, vol 1. Idaho National Laboratory, U.S. Nuclear Regulatory Commission, Office of Nuclear Regulatory Research. NUREG/CR-6903, 2006.

Hallbert, B and A. Kolaczkowski, eds. *The Employment of Empirical Data and Bayesian Methods in Human Reliability Analysis: A Feasibility Study.* Office of Nuclear Regulatory Research, United States Nuclear Regulatory Commission. NUREG/CR-6949. 2007.

Hallbert, B., A. Whaley, R. Boring, P. McCabe and Y. Chang. *Human Event Repository and Analysis (HERA): The HERA Coding Manual and Quality Assurance*, vol 2. Idaho National Laboratory, U.S. Nuclear Regulatory Commission, Office of Nuclear Regulatory Research. NUREG/CR-6903. 2007.

Harding, A. L. and K. B. Byers. "Epidemiology of Laboratory-Associated Infections." In *Biological Safety: Principles and Practices,* Third Edition, 35-56. Fleming, D. O. and D. L. Hunt, eds. Washington D.C.: ASM Press, 2000.

Helmreich, R. L. "On Error Management: Lessons from Aviation." *British Medical Journal*, vol. 320, no. 7237 (2000): 781-785.

Helmreich, R.L., and A. C. Merritt. *Culture at Work in Aviation and Medicine: National, Organizational, and Professional Influences.* Brookfield VT: Ashgate Publishing, 1998.

Kortepeter, M. G., J. W. Martin, J. M. Rusnak, T. J. Cieslak, K. L. Warfield, E. L. Anderson, and M. V. Ranadive. "Managing Potential Laboratory

Exposure to Ebola Virus Using a Patient Biocontainment Care Unit." *Emerging Infectious Diseases.* (2008).

Lentzos, F. "Regulating Biorisk: Developing a Coherent Policy Logic (Part II)." *Biosecurity and Bioterrorism: Biodefense Strategy, Practice, and Science*, vol. 5, no. 1 (2007): 55-61.

Lofstedt, R. "Good and Bad Examples of Siting and Building Biosafety Level 4 Laboratories: A Study of Winnipeg, Galveston and Etobicoke." *Journal of Hazardous Materials*, 93 (2002): 47-66.

Miller, D. and J. Forester. *Aviation Safety Human Reliability Analysis Method (ASHRAM).* Sandia National Laboratories. SAND2000-2955. 2000.

Minnema, D. M. *Improving Safety Culture: Recognizing the Underlying Assumptions.* Powerpoint presentation for the ISM Workshop, Defense Nuclear Facilities Safety Board, 2007.

National Academy of Public Administration for the Federal Aviation Administration. *A Review of the Aviation Safety Reporting System: A Report.* 1994.

Newsletter of the European Biosafety Association. Biosafety Organisation in Spain. *EBSA 2001 Newsletter*, vol. 1, no. 3 (2001).

Paradies, M., L. Unger, P. Haas, and M. Terranova. *Development of the NRC's Human Performance Investigation Process (HPIP).* NUREG/CR5455. System Improvements, Inc. and Concord Associates, Inc.,1993.

Patankar, M. S. and E. J. Sabin. *Safety Culture Transformation in Technical Operations of the Air Traffic Organization: Project Report and Recommendations.* St. Louis, Mo.: Saint Louis University, 2008.

Patankar, M. S. A "Study of Safety Culture at an Aviation Organization." *International Journal of Applied Aviation Studies*, vol. 3, no. 2 (2003): 243-258.

Patankar, M. S., J. P. Brown, and M. D. Treadwell. *Safety Ethics: Cases from Aviation, Healthcare, and Occupational and Environmental Health.* Aldershot, U.K.: Ashgate Publishing, 2005.

Patankar, M. S. and D. Driscoll. "Preliminary Analysis of Aviation Safety Action Programs in Aviation Maintenance." *Proceedings of the First Safety Across High-Consequence Industries Conference, St. Louis, Mo.*, 97-102. 2004.

Patankar, M.S. and J. C. Taylor. *Risk Management and Error Reduction in Aviation Maintenance.* Aldershot, U.K.: Ashgate Publishing, 2004.

Patankar, M. S., T. Bigda-Peyton, E. Sabin, J. Brown, and T. Kelly. *A Comparative Review of Safety Cultures.* St. Louis, Mo.: Saint Louis University, 2005.

Peterson, L.K., E. H. Wight, and M.A. Caruso. "Evaluating Internal Stakeholder Perspectives on Risk-Informed Regulatory Practices for the Nuclear Regulatory Commission." Paper presented at the WM '03 Conference, Tuscon Ariz., 2003.

Pounds, J. and A. Isaac. *Development of an FAA-EUROCONTROL Technique for the Analysis of Human Error in ATM.* DOT/FAA/AM-02/12. Federal Aviation Administration, Office of Aerospace Medicine. 2002.

Race, M. S. "Evaluation of the Public Review Process and Risk Communication at High-Level Biocontainment Laboratories." *Applied Biosafety*, vol. 13, no. 1 (2008): 45-56.

Race, M. S. and E. Hammond. "An Evaluation of the Role and Effectiveness of Institutional Biosafety Committees in Providing Oversight and Security at Biocontainment Labs." *Biosecurity and Bioterrorism: Biodefense Strategy, Practice, and Science*, vol. 6, no. 1 (2008): 19-35.

Reason, J. "Human Error: Models and Management." *British Medical Journal*, vol. 320, no. 7237 (2000): 768-770.

Rusnak, J. M., M.G. Kortepeter, R.J. Hawley, A.O. Anderson, E. Boudreau, and E. Eitzen. "Risk of Occupationally Acquired Illnesses from Biological Threat Agents in Unvaccinated Laboratory Workers." *Biosecurity and Bioterrorism: Biodefense Strategy, Practice, and Science*, vol. 2, no. 4 (2004): 281- 93.

Scarborough, A., L. Bailey and J. Pounds. *Examining ATC Operational Errors Using the Human Factors Analysis and Classification System.* DOT/FAA/AM-05/25. Federal Aviation Administration, Office of Aerospace Medicine. 2005.

Schroeder, D., L. Bailey, J. Pounds, and C. Manning. *A Human Factors Review of the Operational Error Literature.* DOT/FAA/AM-06/21. Federal Aviation Administration, Office of Aerospace Medicine. 2006.

Sexton, J. B., E. J. Thomas, and R. L. Helmreich. "Error, Stress and Teamwork in Medicine and Aviation: Cross-sectional Surveys." *British Medical Journal*, vol. 320, no. 7237 (2000): 745-749.

Shappell, S. and D. Wiegmann. *Developing a Methodology for Assessing Safety Programs Targeting Human Error in Aviation.* DOT/FAA/AM-06/24. Federal Aviation Administration, Office of Aerospace Medicine. 2006.

Shappell, S., C. Detwiler, K. Halcomb, C. Hackworth, A. Boquet, and D. Wiegmann. *Human Error and Commercial Aviation Accidents: A Comprehensive, Fine-Grained Analysis Using HFACS.* DOT/FAA/AM-06/18. Federal Aviation Administration, Office of Aerospace Medicine.

2006. GAO. *NASA: Better Mechanisms Needed for Sharing Lessons Learned.* GAO-02-195. Washington, D.C.: January 30, 2002.

U.S. Nuclear Regulatory Commission. Advisory Committee on Reactor Safeguards. *Review and Evaluation of the Nuclear Regulatory Commission Safety Research Program.* NUREG-1635, vol. 7. 2006.

U.S. Nuclear Regulatory Commission. Division of Risk Analysis and Applications, Office of Nuclear Regulatory Research. *Perspectives Gained From the Individual Plant Examination of External Events (IPEEE) Program.* NUREG-1742, vols. 1-2. 2002.

Wedum, A. G. "Pipetting Hazards in the Special Virus Cancer Program." *Journal of the American Biological Safety Program*, vol. 2, no. 2 (1997): 11-21.

West, D.L., D. R. Twardzik, R. W. McKinney, W. E. Barkley, and A. Hellman. "Identification, Analysis, and Control of Biohazards in Viral Cancer Research." In *Laboratory Safety: Theory and Practice*, 167-223. New York, N.Y.: Academic Press, 1980.

Wiegmann, D. A. and S. A. Shappell. "Human Error Perspectives in Aviation." *International Journal of Aviation Psychology*, vol. 11, no. 4 (2001): 341-357.

End Notes

[1] Report of the Transfederal Task Force on Optimizing Biosafety and Biocontainment Oversight.July, 2009. See Appendix D of the Task Force Report for injury and illness information.

[2] In a review of literature published between 1979 and 1999, Harding and Byers (2000) identified 663 cases of subclinical infections and 1,267 overt infections with 22 deaths. Five deaths were of fetuses aborted as the consequence of a maternal LAI. The authors note the general acknowledgment that these numbers "represent a substantial underestimation of the extent of LAIs." Harding, L. and K. Beyers, "Epidemiology of Laboratory-Associated Infections," in Biological Safety: Principles and Practices, Third Edition (Washington, D.C.: ASM Press, 2000), p. 37.

[3] Morbidity and Mortality Weekly Report (MMWR), vol. 51, no. 07 (Feb. 22, 2002): 141-4.

[4] While risks from radiation, toxic and flammable chemicals, and mechanical and electrical hazards are also present in these labs, for the purposes of this report we are primarily focused on the biological risks.

[5] The National Academy of Sciences defines precursors broadly as "the conditions, events, and sequences that precede and lead up to accidents." This definition includes events that are both internal and external to an organization. Phimister et al., Accident Precursor Analysis and Management: Reducing Technological Risk Through Diligence (Washington, D.C.: National Academies Press, 2004).

[6] Select agents are those biological agents and toxins determined by the CDC and/or APHIS to have the potential to pose a severe threat to public health and safety, animal or plant health, or animal or plant products. See 42 C.F.R. §§ 73.3 & 73.4 (CDC – human and overlap agents); 7 C.F.R. § 331.3 (APHIS – plant); 9 C.F.R. §§ 121.3 & 121.4 (APHIS – animal and overlap agents).

[7] Unless exempted under 42 C.F.R. Part 73, 7 C.F.R. Part 331, or 9 C.F.R. Part 121, an entity or individual may not possess, use, or transfer a select agent or toxin without a certification of registration from the CDC or APHIS. An individual or entity must immediately notify the CDC or APHIS and appropriate federal, state, or local law enforcement agencies upon discovering a theft or loss of a select agent or toxin, and notify the CDC or APHIS upon discovering the release of a select agent or toxin. See 42 C.F.R. § 73.19; 7 C.F.R. § 331.19; 9 C.F.R. § 121.19.

[8] GAO, High-Containment Biosafety Laboratories: Preliminary Observations on the Oversight of the Proliferation of BSL-3 and BSL-4 Laboratories in the United States, GAO-08-108T (Washington, D.C.: Oct. 4, 2007) and High-Containment Laboratories: National Strategy for Oversight Is Needed, GAO-09-574 (Washington, D.C.: Sept. 21, 2009).

[9] Labs not working with select agents can be BSL-1, 2, or 3. Some examples of nonselect agents include the micro-organisms that cause HIV, tuberculosis, and typhoid fever.

[10] H.R. 1225, 111th Cong. § 203 (2009); S. 485, 111th Cong. (2009).

[11] Such an approach—in particular, learning from the experiences of other industries—was recommended in the Report of the Transfederal Task Force on Optimizing Biosafety and Biocontainment Oversight.

[12] These agencies include the CDC, NIH, USDA, VA, the Food and Drug Administration (FDA), the Department of Commerce (DOC), the Department of Defense (DOD), the Department of Labor's (DOL) OSHA, the Department of State (State), the Department of Justice's (DOJ) Federal Bureau of Investigation (FBI), the Department of Homeland Security (DHS), the Department of Energy (DOE), the Department of the Interior (DOI), and the Environmental Protection Agency (EPA).

[13] An entity is defined in the select agent regulations as any government agency (federal, state or local), academic institution, corporation, company, partnership, society, association, firm, sole proprietorship, or other legal body. A registered entity may operate multiple labs within a single facility. 42 C.F.R. § 73.1; 7 C.F.R. § 331.1; 9 C.F.R. § 121.1.

[14] The Secretary of HHS developed the Select Agent Program in the CDC in response to the Antiterrorism and Effective Death Penalty Act of 1996. The Public Health Security and Bioterrorism Preparedness and Response Act of 2002 revised and expanded the Select Agent Program within the CDC and granted comparable authority to regulate select agents and toxins affecting plants and animals to the Secretary of Agriculture, a responsibility then delegated to APHIS.

[15] GAO-08-108T and GAO-09-574.

[16] Biosafety in Microbiological and Biomedical Laboratories, Fifth Edition.

[17] J. Reason in S.C. Schleiffer, "We Need to Know What We Don't Know," International Air Safety Seminar, 35 (2005): 333-340.

[18] C.A. Hart, "Stuck on a Plateau: A Common Problem," in Accident Precursor Analysis and Management: Reducing Technological Risk Through Diligence, James R. Phimister, Vicki M. Bier, Howard C. Kunreuther, eds. (Washington, D.C.: National Academies Press, 2004), 151.

[19] J. Reason, "Human Error: Models and Management," British Medical Journal, vol. 320 (2000): 768.

[20] Barach and Small, "Reporting and Preventing Medical Mishaps: Lessons from Non-medical Near Miss Reporting Systems," British Medical Journal, vol. 320 (2000): 759- 763.

[21] Gronvall et al., "High-Containment Biodefense Research Laboratories: Meeting Report and Center Recommendations," Biosecurity and Bioterrorism: Biodefense Strategy, Practice, and Science, vol. 5, no. 1 (2007).

[22] A bibliography of articles used to develop SRS lessons from the literature is available at the end of this report.

[23] GAO, Organizational Culture: Techniques Companies Use to Perpetuate or Change Beliefs and Values, GAO/NSAID-92-105 (Washington, D.C.: Feb. 27, 1992).

[24] GAO, Nuclear Safety: Convention on Nuclear Safety is Viewed by Most Member Countries as Strengthening Safety Worldwide, GAO-10-489 (Washington, D.C.: Apr. 29, 2010).

[25] Reason, "Human Error: Models and Management," 768-770.

[26] See more about the three key areas of SRS design in a review of the literature in the previous section of this report: Program Goals and Organizational Culture Guide Safety Reporting System Design and Implementation in Three Key Areas.

[27] While we collected information on a wide variety of safety reporting programs and systems in the three industries—and in some cases comment on these different programs—we primarily developed our lessons from one reporting program in each of the three industries. We chose to focus on these programs because they represent fairly longstanding, non-regulatory, domestic, industrywide or servicewide reporting programs.

[28] Lucian L. Leape et al., "Promoting Patient Safety by Preventing Medical Error," Journal of the American Medical Association, vol. 280, no.16 (Oct. 28, 1998): 1444-47.

[29] National Transportation Safety Board, Aircraft Accident Report—Transworld Airlines, Inc. Boeing 727-231. NTSB-AAR-75-16 (Washington, D.C., 1975.)

[30] According to the NRC, their Allegation Program evaluates a broad range of nuclear safety concerns associated with NRC-regulated activities, including, for example, complaints of retaliation for raising nuclear safety concerns.

[31] Joseph V. Rees, Hostages of Each Other: The Transformation of Nuclear Safety Since Three Mile Island (Chicago, Ill.: The University of Chicago Press, 1994).

[32] Rees, Hostages of Each Other.

[33] The Davis-Besse nuclear power plant in Ohio was shut down between 2002 and 2004 because leakage had caused extensive corrosion on the vessel head—a vital barrier preventing a radioactive release. Significant to the failure and to the delay in restarting the plant were NRC's concerns over the plant's safety culture. GAO, Nuclear Regulation: NRC Needs to More Aggressively and Comprehensively Resolve Issues Related to the Davis-Besse Nuclear Power Plant's Shutdown, GAO-04-415 (Washington, D.C.: May 17, 2004).

[34] The PSRS was discontinued at the end of fiscal year 2009. We include the PSRS in our case study with the PSIS because it was central to the design of VA's safety reporting program and it operated for nearly 10 years, providing valuable insights for SRS lessons learned.

[35] L. Leape et al., "Promoting Patient Safety by Preventing Medical Error."

[36] National Academy of Public Administration, A Review of The Aviation Safety Reporting System (1994).

[37] GAO, Aviation Safety: Improved Data Quality and Analysis Capabilities Are Needed as FAA Plans a Risk-Based Approach to Safety Oversight, GAO-10-414 (May 6, 2010). The FAA runs a number of safety reporting systems, several of which are reviewed in this recent GAO report. See also American Institutes for Research, Best Practices for Event Review Committees (December 2009): 1-2.

[38] Despite the increase in the overall number of reports, the proportion of serious reports has declined over the years. Rather than suggesting an increase in safety problems, the increasing number of reports—especially those at the lower half of the risk pyramid—indicates a robust reporting culture, where workers are more aware of and willing to report safety issues at the incident or concern level.

[39] INPO afforded us substantial access to their liaison. In multiple interviews over the period of the investigation, the liaison explained details of INPO history and policy that are not widely available because of the centrality of confidentiality to INPO's safety operations from its initiation. We confirmed these details, when possible, from documents. The facts we report were further vetted by an official INPO spokesman. We explain INPO's confidentiality efforts later in this report.

[40] GAO, VA Patient Safety: Initiatives Promising, but Continued Progress Requires Culture Change, T-HEHS-00-167 (Washington, D.C.: July 27, 2000).

[41] GAO, VA Patient Safety Program: A Cultural Perspective at Four Medical Facilities, GAO-05-83 (Washington, D.C.: Dec. 22, 2004).

[42] W.D. Reynard, C.E. Billings, E.S. Cheaney and R. Hardy, The Development of the NASA Aviation Safety Reporting System, Pub 34, NASA Reference Publication (1986): 25.

[43] These are known as "enforcement incentives" inside the FAA.

[44] American Institutes for Research, Best Practices for Event Review Committees. 45GAO-05-83.

[46] INPO has specifically defined the criteria for reports "noteworthy" enough that they should be sent on to INPO for central analysis. The criteria include events that caused an unexpected change in conditions or had the potential to cause these changes under slightly different circumstances.

[47] In terms of the risk pyramid, the VA SRS programs expanded reporting from top-level events (accidents) to include midlevel events (incidents).

[48] J.M. Beaubien and D. P. Baker, "A Review of Selected Aviation Human Factors Taxonomies, Accident/Incident Reporting Systems, and Data Reporting Tools," International Journal of Applied Aviation Studies, vol. 2, no. 2 (2002); M. Tamuz and E. J. Thomas, "Classifying and Interpreting Threats to Patient Safety in Hospitals: Insights from Aviation," Journal of Organizational Behavior, 27 (2006): 919-940.

[49] In September 2009, the AHRQ published for review a follow-up version to its 2008 Common Formats for adverse medical events, required by the Patient Safety and Quality Improvement Act of 2005. The process of developing these codes stretched over 3 to 4 years.

[50] GAO-10-414.

[51] Academy of Engineering, Accident Precursor Analysis and Management: Reducing Technological Risk through Diligence (Washington, D.C.: National Academies of Science, 2004): 14.

[52] Several aviation SRSs in other countries have suffered from perceptions they failed to maintain the confidentiality of reporters or from lack of funding. The Canadian Securitas—responsible for receiving safety reports from aviation, rail, and marine industries—is so under-resourced that its budget supports less than one employees per province. The original aviation reporting system in New Zealand failed due to a breach of confidentiality. An Australian aviation reporting system that had functioned for many years was weakened under social pressures for redress and pressure from the regulator after a fatal aviation accident. Those pressures resulted in an indirect breach of identity and a change in the law

toward "natural justice" for reporters. A representative of the Australian SRS reported in 2008 that the number of reports to the SRS had fallen.

[53] 14 C.F.R. § 91.25. If the incident is found to involve a violation of regulations, neither civil penalties nor certificate suspension will be imposed as long as the reported action (1) is not deliberate and (2) does not involve a criminal offense, accident, or evidence of incompetence, and the reporter (1) has not been in violation for 5 years and (2) completed and submitted a report under ASRS within 10 days of the incident. Advisory circular AC-00-46D.

[54] Several other voluntary SRS programs, such as ASAPs, stress corrective actions over punishment, although the FAA can prosecute cases involving egregious acts (e.g., substance or alcohol abuse or the intentional falsification of information). ASAPs provide previously unavailable information rapidly and directly from those responsible for day-to-day aviation operations. While the FAA has limited access to ASAP data, these programs are expected to lead to improvements in aviation safety.

[55] GAO-10-414.

[56] This policy of protecting INPO reports from public disclosure was tested by a request under the Freedom of Information Act (FOIA) for INPO safety reports that had been provided to the NRC. In Critical Mass Energy Project v. Nuclear Regulatory Commission, the U.S. Court of Appeals for the District of Columbia upheld the lower court decision that information voluntarily provided by INPO to the NRC, which was commercial in nature and not customarily released to the public, was confidential and therefore exempt from disclosure under FOIA. 975 F.2d 871 (D.C. Cir. 1992), cert. denied, 507 U.S. 984 (1993).

[57] The NRC also runs a reporting system—the Allegations Program—for nuclear safety or regulatory concerns involving NRC regulated facilities and licensed nuclear material. For this program, there are exceptions to FOIA and related regulations that may justify withholding information that would identify an alleger or other confidential source. See 5 U.S.C. §§ 552(b)(6), (7); 10 C.F.R. §§ 9.17(a)(6), (7). Confidentiality is not routinely offered; however, when reporters request it, it is formalized in a letter that establishes several conditions under which confidentiality will not be preserved, such as a request from Congress or state or federal law enforcement bodies.

[58] Rees, Hostages of Each Other.

[59] Pub. L. No. 109-41, 119 Stat. 424 (July 29, 2005).

[60] GAO, Patient Safety Act: HHS Is in the Process of Implementing the Act, So Its Effectiveness Cannot Yet Be Evaluated, GAO-10-281 (Washington, D.C.: Jan. 29, 2010).

[61] National Academy of Public Administration, A Review of The Aviation Safety Reporting System.

[62] GAO, Nuclear Safety: Convention on Nuclear Safety Viewed by Most Member Countries as Strengthening Safety Worldwide, GAO-10-489 (Washington, D.C.: Apr. 29, 2010).

[63] There are four major parts of the inspection review process: (1) performance indicators, (2) analysis of corrective action reports (data mining that looks for word trending), (3) plant evaluation process (on-site interviews with a variety of staff areas and levels), and (4) safety culture surveys.

[64] See the AHRQ Web site, www.ahrq.gov/about/psimpcorps.htm.

[65] Rand Corporation, Evaluation of the Patient Safety Improvement Corps: Experiences of the First Two Groups of Trainees (2006).

[66] P.D. Mills, J. Neily, L.M. Kinney, J. Bagian, W.B. Weeks, "Effective Interventions and Implementation Strategies to Reduce Adverse Drug Events in the Veterans Affairs (VA) System," Quality and Safety in Health Care, 17 (2008): 37-46.

[67] While we sometimes refer to the agencies generally, this section specifically applies to the CDC and APHIS Select Agent Program.

[68] Under the Select Agent Regulations, individuals or entities must immediately notify the CDC or APHIS and appropriate federal, state, or local law enforcement agencies upon discovering a theft or loss of a select agent or toxin, and notify the CDC or APHIS upon discovering the release of a select agent or toxin. See 42 C.F.R. § 73.19; 7 C.F.R. § 331.19; 9 C.F.R. § 121.19. The individual or entity that discovered the theft, loss, or release must submit an APHIS/CDC Form 3 (Report of Theft, Loss, or Release of Select Agents and Toxins) within 7 calendar days.

[69] For example, Patankar et al. note that, "There are three key issues regarding research and measurement of safety culture: (a) Survey instruments take a 'snapshot' measurement of safety climate. When such measurements are repeated across multiple organizational units and conducted repeatedly over a reasonably long time (over five years), a cultural assessment can be developed. (b) A rigorous analysis of the various factors that influence safety climate/culture needs to be conducted so as to better understand the interrelationship among these factors and their individual, group, and cumulative influence on the overall safety climate/culture.... (c) Results from measurements need to be distributed consistently throughout the organization so that everyone is fully aware of their contributions to the goals and are able to make timely actions/changes that are consistent with the organizational goals." M.S. Patankar, T. Bigda-Peyton, E. Sabin, J. Brown, and T. Kelly, A Comparative Review of Safety Cultures (St. Louis, Mo.: Saint Louis University, 2005).

[70] GAO, VA Patient Safety Program: A Cultural Perspective at Four Medical Facilities, GAO-05-83 (Washington, D.C.: Dec. 15, 2004); and R.L. Helmreich and A.C. Merritt, Culture at Work in Aviation and Medicine: National, Organizational, and Professional Influences (Brookfield VT, U.K.: Ashgate Publishing, 1998).

[71] National Nuclear Security Administration, Lessons Learned and Recommendations from Review of NASA's Columbia Accident Investigation Board Report (2004).

[72] S.M. Evans et al., "Attitudes and Barriers to Incident Reporting: A Collaborative Hospital Study," Quality and Safety in Health Care, 15 (2006): 39-43; and Tamuz, M. and E. J Thomas, "Classifying and interpreting threats to patient safety in hospitals: insights from aviation," Journal of Organizational Behavior, 27 (2006): 919-940.

[73] B.J. Weiner, C. Hobgood, and M. Lewis, "The Meaning of Justice in Safety Incident Reporting," Social Science & Medicine, vol. 66, no. 2 (2008): 403-413.

[74] Also available at http://www.selectagents.gov/resources/CDCAPHIS_ Theft_Loss_ Release_ Information_Document.pdf

[75] The literature and case studies also suggest that reporting increases do not necessarily signal an increase in safety problems, but rather an increased awareness of reportable incidents and trust in the reporting system.

[76] This may include reasonably anticipated skin, eye, mucous membrane, or parenteral contact with blood or other potential infectious materials that may result from the performance of a person's duties.

[77] A.L. Harding and K. B. Byers, "Epidemiology of Laboratory-Associated Infections," in Biological Safety: Principles and Practices, Third Edition, D.O. Fleming and D. L. Hunt, eds. (Washington D.C.: ASM Press, 2000), 35-56.

[78] BD Biosciences' Aerosol Management Option (AMO) system, Model 333728 (US) and 333729 (Europe).

[79] 42 U.S.C. § 262a(h) (Disclosure of information).

[80] In April 2010, the labs were provided a confidential means, by the agencies, for reporting safety and security issues associated with the possession, use, and transfer of select agents and toxins. HHS's Office of Inspector General maintains a hotline that allows individuals to anonymously report fraud, waste, and abuse in all departmental programs. This hotline is now available to anonymously report safety or security issues related to select agents and toxins.

[81] Entities that continue to have repeat noncompliance of the Select Agent regulations can be placed on a Performance Improvement Plan. Entities can also be referred to the Office of Inspector General (OIG) for Select Agent violations, which can result in civil monetary penalties.

[82] (1) Administrative actions: The CDC and APHIS may deny an application or suspend or revoke a registered entity's certificate of registration. (2) Civil money penalties or criminal enforcement: the CDC refers possible violations of the select agent regulations to the HHS's Office of Inspector General (OIG). The HHS-OIG can levy civil money penalties (for an individual, up to $250,000 for each violation and, for an entity, up to $500,000 for each violation) or recommend criminal enforcement (imprisonment for up to 5 years, a fine, or both). APHIS relies on its own investigative unit, USDA Marketing and Regulatory Programs—Investigative and Enforcement Services (IES), for initial investigations of potential select agent violations. Like the HHS-OIG, IES can levy civil money penalties or recommend criminal enforcement. IES refers potential criminal violations to USDA's OIG. (3) Referral to the Department of Justice: the CDC or APHIS can refer possible criminal violations involving select agents to the department for further investigation or prosecution.

[83] 14 C.F.R. § 91.25. If the incident is found to involve a violation of regulations, neither civil penalties nor certificate suspension will be imposed as long as (1) the reported action is not deliberate and does not involve a criminal offense, accident, or evidence of incompetence and (2) the reporter has not been in violation for 5 years and completed and submitted a report under ASRS within 10 days of the incident. FAA Advisory Circular 00- 46D.

[84] GAO, Results-Oriented Government: GPRA Has Established a Solid Foundation for Achieving Greater Results, GAO-04-38 (Washington, D.C.: Mar. 10, 1994).

[85] See appendix II for a summary of lessons derived from the literature and case studies that can be applied to an SRS for biological labs.

[86] The total number and locations of all biological laboratories is unknown, and, as a result, in a 2009 report (GAO-09-574), we recommended that a process to identify them be initiated. In addition, there is no centralized oversight responsibility for labs except for those registered with the Select Agent Program. Lab safety is generally covered through the OSHA or state regulations for general organizational safety. The principles of biosafety and biocontainment have been articulated in two key documents, the NIH Guidelines for Research Involving Recombinant DNA Molecules (NIH Guidelines) and the CDC-NIH manual, Biosafety in Microbiological and Biomedical Laboratories. Research that involves recombinant DNA molecules may be subject to the NIH Guidelines. Compliance with the NIH Guidelines is a term and condition of NIH grants and thus is mandatory for all institutions that receive NIH funding for recombinant DNA research. In addition, a number of other federal agencies (e.g., the Department of Energy, Department of the Army, USDA, and VA to name a few) have made compliance with the NIH Guidelines a term and condition of research grants and a requirement for their own intramural research activities. Although adherence to the BMBL is voluntary, the manual is a widely accepted code of practice for biosafety and biocontainment in all microbiological and biomedical laboratories in the United States and in many other countries.

[87] While entity-specific information is protected from release under FOIA, after the CDC provided the data in response to a congressional request, specific entity information was somehow leaked to the media.

[88] Harding, L. and K. Beyers, "Epidemiology of Laboratory-Associated Infections," in Biological Safety: Principles and Practices, Third Edition (Washington, D.C.: ASM Press, 2000), p. 37.

In: Biological Laboratories
Editors: J. Hopkins and N. Caldwell
ISBN: 978-1-62257-822-1

Chapter 2

OVERSIGHT OF HIGH-CONTAINMENT BIOLOGICAL LABORATORIES: ISSUES FOR CONGRESS*

Frank Gottron and Dana A. Shea

SUMMARY

The federal government responded to the September 11, 2001, terrorist attacks and the subsequent anthrax attacks with increased focus on and funding for biodefense. A key consideration in this response was addressing shortages in diagnostic, clinical, and research laboratory capacity. Several departments and agencies have increased or are in the process of increasing their laboratory capacity. High-containment laboratories play a critical role in the biodefense effort, offering the hope of better responses to an attack and a better understanding of the threat posed by bioterrorism. However, they also could increase the risk of a biological attack by serving as a potential source of materials or training. Indeed, the Commission on the Prevention of Weapons of Mass Destruction Proliferation and Terrorism recommends tightening government oversight of high-containment laboratories.

Policymakers have become increasingly interested in the oversight of these facilities following reports of accidents, regulatory noncompliance, and

---

* This is an edited, reformatted and augmented version of a Congressional Research Service publication, CRS Report for Congress R40418, dated May 4, 2009.

community resistance. The increase in high-containment laboratory capacity has raised new policy questions and emphasized existing ones. How much laboratory capacity is enough? What is the necessary federal investment? Should laboratories be consolidated or dispersed? What plans exist to coordinate multiple agency efforts to expand high-containment laboratory capacity? Does increasing laboratory capacity increase the risk of accidents and the opportunity for purposeful misuse? What is an acceptable balance between the benefits these laboratories provide and the risks they pose?

Interested Members of Congress might take action to address some or all of these concerns. Alternatively, they might defer action until efforts currently under way assess and make recommendations regarding the existing regulatory structure. If Congress chooses to enhance oversight, it might require a survey of existing facilities and their use and a national needs assessment, perhaps barring further construction until these are complete. Stakeholders could focus on enhancing self-regulatory activities such as improving or standardizing laboratory worker training or building a mechanism for sharing lessons learned. Rather than relying on self- regulation, policymakers might enhance oversight through additional regulation of high-containment facilities, requiring laboratory or personnel certification, or by broadening the Select Agent Program. Which agencies should implement any new mandates remains an open question.

Biocontainment technologies are widely used by scientists around the world. Efforts to increase control of U.S. high-containment laboratories may put domestic industry at a competitive disadvantage and inhibit international academic collaboration. Absent international harmonization, the United States can only partially address the threat of a high-containment laboratory being the source of a bioterror weapon.

A key task for policymakers is to define their goals for enhancing oversight of high-containment laboratories. The focus of the oversight effort may affect which policy issues are addressed. For example, focusing on a registry of existing high-containment laboratory capacity may improve planning, coordination, and efficiency of use but provide relatively limited security benefits. Similarly, a rigorous oversight program including facility and personnel licensure, mandatory training, and restricted construction of new facilities may provide security benefits at the cost of regulatory burden, increased federal expenditures, and impeded scientific progress in countermeasure research, bioforensics, and public health. When weighing options to address these complex policy issues, policymakers may have to reconcile many competing and potentially conflicting national needs.

INTRODUCTION

The federal government responded to the September 11, 2001, terrorist attacks and the subsequent anthrax attacks with increased focus on and funding for biodefense. A key consideration in this response was addressing capacity shortages for diagnostic, clinical, and research laboratories. Since 2001, the Department of Defense (DOD), the Department of Homeland Security (DHS), the Department of Health and Human Services (HHS), and the Department of Agriculture (USDA) have increased or are in the process of increasing their laboratory capacity for the study of dangerous pathogens. Because stringent protocols and engineering are used to contain the most dangerous pathogens, these facilities are often referred to as high-containment laboratories. High-containment laboratories play a critical role in the biodefense effort, offering the hope of better responses to a biological attack and a better understanding of the bioterrorism threat. However, they could also increase the risk of a biological attack by being a source of materials or training.

In addition to increasing laboratory capacity, the federal government invested in research to develop countermeasures, diagnostics, and detectors for dangerous pathogens potentially usable in a terrorist attack. One group has calculated that congressionally appropriated funding for civilian biodefense increased from $690 million in FY2001 to $5.4 billion in FY2008.[1] Much of this funding supports academic and industrial researchers and laboratories that handle these dangerous pathogens. The influx of funds and researchers has exacerbated existing concerns about aging laboratory infrastructure and limited research and diagnostic laboratory capacity.

Non-federal entities have also expanded or constructed additional high-containment laboratories. In addition to the threat of bioterrorism, an increasing awareness of the threat posed by emerging and re-emerging diseases has led to the proliferation of high-containment laboratories internationally, as the technologies used are widely available.

The increase in high-containment laboratory capacity has raised new policy questions and increased focus on existing ones. How much laboratory capacity is enough? What is the necessary federal investment? Should laboratories be consolidated or dispersed? What is the optimal size and type of high-containment laboratory? With multiple agencies expanding high-containment laboratory capacity, is a plan coordinating these efforts necessary? Does increasing laboratory capacity and the number of trained scientists increase the risk of accidents and/or opportunities for purposeful misuse? What is an acceptable balance between the benefits these laboratories provide and the risks they pose?

Policymakers have become increasingly interested in the expansion of these facilities following reports of accidents, regulatory noncompliance, and recent examples of community resistance to the laboratories. The Commission on the Prevention of Weapons of Mass Destruction Proliferation and Terrorism recommended tightening government oversight of high-containment laboratories.[2] These laboratories may also be regulated by state and local laws, regulations, and ordinances. This chapter focuses on the federal government's regulation of high containment laboratories. It explains the concepts of biosecurity and biosafety, including the mechanisms by which the federal government oversees their implementation; describes the proliferation of high-containment laboratories; discusses issues facing federal policymakers; and identifies policy options for congressional consideration.

BIOSECURITY AND BIOSAFETY

Biosecurity and biosafety are closely related. Their definitions vary from source to source and sometimes overlap. In this chapter, biosecurity refers to steps taken to secure pathogens or biological materials from theft, unauthorized access, or illegal use, while biosafety refers to mechanisms or practices employed to lower the risk of unintentional infection in the laboratory or environmental release from the laboratory.[3] While biosecurity and biosafety are clearly related closely, their practices and oversight mechanisms have mainly developed independently.

Biosecurity and the Select Agent Program

Although biosafety has been a concern of the general laboratory science community for many years, biosecurity has only recently come to the fore. The major federal regulatory program addressing biosecurity is the Select Agent Program.[4] The federal government created the Select Agent Program in order to regulate and oversee commerce in pathogens that might have severe consequences if released into the environment.

The Select Agent Program was first established by the Antiterrorism and Effective Death Penalty Act of 1996 (P.L. 104-132). This law required HHS to identify a list of organisms and toxins that could potentially be used for bioterrorist attacks and to regulate their transfer, though not their possession. These pathogens and toxins were to be known as select agents.

The 2001 anthrax mailings increased the federal government's interest in the threat of bioterrorism. Congress enacted the USA PATRIOT Act (P.L. 107-56) and the Public Health Security and Bioterrorism Preparedness and Response Act of 2002 (P.L. 107-188) which increased restrictions on the possession of pathogens and toxins that have been identified as possessing the greatest potential for malevolent use. Entities possessing select agents were required to develop explicit security and biosafety plans and procedures to be reviewed and certified by the federal government.[5] The Select Agent Program oversees regulation of select agents possession, transfer, and use. It focuses mainly on two areas: people who have access to select agents and facilities where select agents are used and stored.

Scientists who want access to select agents must register with the federal government. Applicants provide information to the Department of Justice (DOJ), which performs background checks to determine whether the applicant may be permitted to handle select agents. Such permits are nontransferable and are valid for five years.

Facilities that possess or use select agents must develop and implement a security plan to protect the select agents from theft or improper access. These plans are submitted either to HHS, through the Centers for Disease Control and Prevention (CDC), or to USDA, through the Animal and Plant Health Inspection Service (APHIS). These agencies review them for regulatory compliance and audit their security plan implementation. The CDC performs announced inspections of each regulated facility at least once every three years.[6]

Failure to comply with the regulations of the Select Agent Program is punishable by fines. Persons possessing select agents who have not successfully completed the DOJ security risk assessment process or are prohibited from possessing select agents under the USA PATRIOT Act may be in violation of 18 U.S.C. 175, which prohibits the possession of biological weapons.

Program Participation

As of February 2009, approximately 390 entities had been issued certificates allowing work with select agents and 15,300 staff had active security risk assessment approvals to have access to select agents.[7] See **Table 1** for details. The largest group of certified entities are owned or operated by state or local governments. Many of these entities are state or local public health laboratories. As part of efforts to develop robust testing and response capabilities following a biological attack, the federal government has developed a Laboratory Response Network that links many of these state or local public health laboratories. One of the goals in establishing the Laboratory Response Network was to create a laboratory in each state capable of handling dangerous pathogens.[8]

Additionally, the large number of academic entities, and presumably researchers, has increased tensions over the Select Agent Program in academia. This community has a long tradition of openness and has, in some cases, viewed the select agent regulations as too onerous to continue performing research with these pathogens. Approximately 200 entities either transferred or destroyed their select agent inventories rather than registering under the select agent regulations.[9]

Table 1. Entities That Work with Select Agents

Type of Entity	CDC Regulated	USDA Regulated	Total
Academic	95	22	117
Government—State/Local	117	10	127
Government—Federal	62	9	71
Private	49	22	71
Total	323	63	386

Source: Centers for Disease Control and Prevention, personal communication, February 9, 2009; and Department of Agriculture, personal communication, February 4, 2009.

Program Oversight

The performance of the Select Agent Program has been the topic of a number of investigations by agency Inspectors General (IG). These investigations have focused on agency internal controls on select agents,[10] the ability of lead agencies to process researcher and entity applications,[11] and the regulatory compliance of facilities receiving select agent certificates. The HHS IG found that 11 of 15 representative universities investigated did not fully comply with the select agent regulations.[12] It also found that none of the eight representative state, local, private, or commercial laboratories investigated were in full compliance. Three of these entities chose to withdraw their select agent licenses following the investigation.[13] The USDA IG found similarly that APHIS controls over registered entities and registered entities' compliance with select agent regulations were not complete.[14] In 2007, the CDC suspended a select agent entity certificate at Texas A&M University, reportedly because of failure to report occupational exposures in a timely fashion and because laboratory workers lacking a security risk assessment were allowed access to select agents.[15] Other companies, laboratories, and universities have also been cited and fined for violations of the select agent regulations. The HHS OIG has levied a total of $1,887,000 in fines on 12 organizations for failure to comply with the Select Agent regulations.[16]

Biosafety and Laboratory Best Practices

Scientists studying dangerous pathogens generally acknowledge the risk of contracting the disease under study and the potential of accidentally releasing a pathogen into the environment. Over time, scientists have developed best practices to mitigate these risks. In the 1970s and 1980s, the U.S. government began collecting these practices and issuing them as formal guidelines. The most commonly referenced guidelines are disseminated by HHS in the publication *Biosafety in Microbiological and Biomedical Laboratories* (BMBL).[17]

Biosafety Levels

The Centers for Disease Control and Prevention (CDC) and the National Institutes of Health (NIH) have established four main levels of biosafety (BSL-1 to BSL-4) to guide laboratory researchers in the safe handling of biological agents.[18] Each biosafety level is associated with specific physical and procedural protections. See **Table 2**. In general, the more dangerous the pathogen is to human health, the higher its recommended biosafety level. Each successive biosafety level consists of the protective measures of the lower biosafety levels, augmented with additional protective measures. Generally, the term "high-containment laboratory" refers to BSL3 and BSL-4 laboratories. The current BMBL guidelines grew out of earlier efforts by the DOD, HHS, and USDA. These best practices were developed to protect laboratory workers based on the various specific risks posed by different pathogens.[19] The guidelines present criteria to define the different biosafety levels, best practices for safe laboratory operation at these biosafety levels, and mechanisms for identifying the appropriate biosafety level for a given organism and procedure. Because the biosafety level may vary depending on the procedure, some experts recommend that biosafety levels be thought of as linked to a combination of pathogen and experiment, not as accompanying particular pathogens.[20]

Table 2. Biosafety Level Criteria

BSL	Agents	Practices	Primary Barriers and Safety Equipment	Facilities
1	Not known to cause disease in healthy humans	Standard microbiological practices	None required	Open bench and sink required
2	Agents associated with human disease Routes of transmission include percutaneous injury, ingestion, and mucous membrane exposure	BSL-1 practice plus: Limited access Biohazard warning signs Sharp objects precautions Biosafety manual defining waste decontamination and medical surveillance policies	Class I or II Biological Safety Cabinets (BSCs) or other physical containment devices used for all manipu-lations of agents that cause splashes or aerosols of infectious materials Laboratory coats and gloves required Face protection as needed	BSL-1 plus: Autoclave available
3	Indigenous or exotic agents with potential for aerosol transmission Disease may have serious or lethal consequences	BSL-2 practice plus: Controlled access All waste decontaminated Laboratory clothing decontaminated before laundering Baseline worker health established	Class I or II BSCs or other physical containment devices used for all open manipulation of agents Protective laboratory clothing and gloves equired Respiratory protection as needed	BSL-2 plus: Physical separation from access corridors Self-closing, double-door access Exhaust air not recirculated Negative airflow into laboratory
4	Dangerous/exotic agents which pose high risk of life-threatening disease and for which aerosol-transmitted laboratory infections have occurred Related agents with unknown risk of transmission	BSL-3 practices plus: Clothing change before entering Shower on exit All material decontaminated on exit from facility	All procedures conducted in Class III BSCs or Class I or II BSCs in combination with full-body, air-supplied, positive pressure personnel suit	BSL-3 plus: Separate building or isolated zone Dedicated supply and exhaust, vacuum, and decontamination systems Other requirements outlined in the BMBL

Source: BMBL, 5th edition, 2007.

Table 3. Recommended Biosafety Levels for Some Select Agents

Pathogen (Disease)	Recommended BSL
Bacillus anthracis (anthrax)	2 or 3
Ebola virus	4
Marburg virus	4
Francisella tularensis (tularemia)	2 or 3
Variola major (smallpox)	4[a]
Yersina pestis (plague)	2 or 3
Ricin toxin	2 or 3

Source: BMBL, 5th edition, 2007.

a. An international agreement limits research with live smallpox virus to two specific laboratories, one in the United States and the other in Russia.

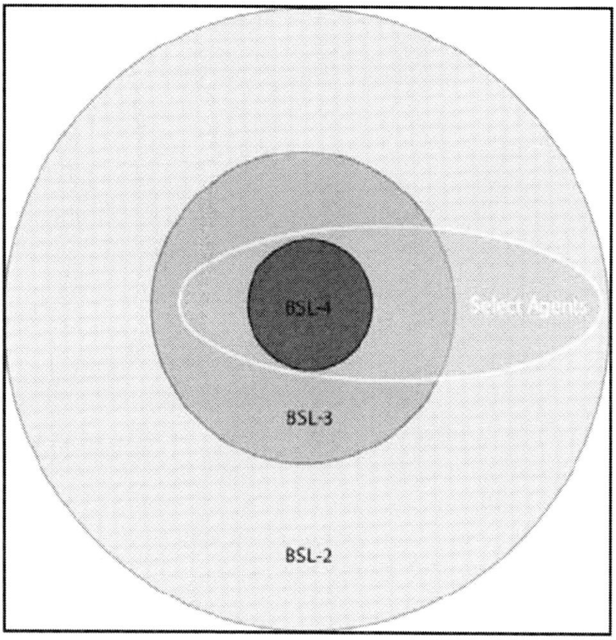

Source: CRS analysis of BMBL and select agent list.

Notes: Areas are for illustrative purposes only and are not to scale. Although some BSL-4 work does not use Select Agents, all U.S. BSL-4 facilities work with some Select Agents. Thus, all U.S. BSL-4 facilities are regulated under the Select Agent Program.

Figure 1. Correlation of Regulation by the Select Agent Program with Biosafety Level.

The BMBL describes primary and secondary barriers to infection, as well as standard and special laboratory practices for each biosafety level. Unlike previous editions, the most recent edition of the BMBL also contains some security guidelines for the different biosafety levels.[21] According to federal grant policy and standard contract language, researchers and laboratory workers at institutions receiving federal funds are to be trained in the procedures described in the BMBL before they gain access to the laboratory.

The correct biosafety level depends on the pathogen, its potential for transmission and treatment, and the research procedure or activity being conducted. Procedures deemed unlikely to produce disease in healthy humans should be conducted at BSL-1. Those that may cause disease in healthy humans, but for which immunization or antibiotic treatment is available, should be conducted at BSL-2. Procedures that may cause serious or potentially lethal diseases as a result of pathogen inhalation should be conducted at BSL-3. Procedures that pose a high individual risk of aerosol-transmitted laboratory infections and life-threatening disease should be conducted at BSL-4.[22]

Because the recommended biosafety level for a particular activity is determined by considering both the pathogen and procedure, it is possible to safely handle potentially deadly pathogens at lower biosafety levels when performing certain types of experiments or activities. Several pathogens thought to pose the greatest health threat, as defined by their inclusion on the select agent list, may be handled outside high-containment facilities. See **Table 3**. For example, the BMBL recommends the bacteria that causes the disease anthrax be handled at BSL-2 for "activities using clinical materials and diagnostic quantities of infectious cultures" and at BSL-3 for "work involving production quantities or high concentrations of cultures, screening environmental samples (especially powders) from anthrax-contaminated locations, and for activities with a high potential for aerosol production."[23] Of the 73 human and animal select agents and toxins, 13 warrant BSL-4 containment for any procedure. However, 30 others may be studied at BSL-2 containment depending on the circumstances.[24]

Some of the work performed in BSL-4 facilities is on pathogens that are not select agents.[25] All of the BSL-4 facilities in the United States perform at least some work with select agents and therefore are regulated under the Select Agent Program. Only some of the work in BSL-3 facilities uses select agents, so many BSL-3 facilities are not subject to select agent regulations. A relatively small proportion of BSL-2 facilities fall under the Select Agent Program. See **Figure 1**.

Guideline Enforcement

The BMBL guidelines are best practices. Compliance with these guidelines is typically voluntary, but it is widely adopted as facility policy. Some mandatory adoption and compliance requirements exist. For example, federal contracts and grants require compliance with the BMBL guidelines, and noncompliance can result in loss of current and potentially future federal funding. Some federal agencies have promulgated management directives requiring compliance with the BMBL in agency laboratories. Industrial, academic, and non-profit laboratories generally establish health and safety officials or other responsible persons whose responsibilities include verifying the good operation of protective equipment installed in laboratories and thus continued compliance with the guidelines. Finally, laboratories regulated under the Select Agent Program are required to consider the BMBL guidelines when preparing the biosafety component of required security plans.

In addition to voluntary compliance, the BMBL guidelines are incorporated into building design standards, grant and contract requirements, other federal requirements related to DNA research, and the Select Agent Program.

Building Design Standards

Federal facilities, and most other facilities built using federal funds, are built to design standards established by the agency that builds a facility or funds its construction. These design standards generally adhere to the guidelines promulgated in the BMBL. The design standards may contain or refer to other security requirements.

For example, the National Institutes of Health has design specifications for NIH buildings. These specifications provide requirements for systems such as ventilation, utilities, and access controls.[26] Security requirements are specified in a separate document.[27] For high-containment laboratories, NIH issues checklists to ensure that requirements are tested and determined to be satisfactory before the laboratory is occupied.[28] This testing process is often called certifying or commissioning a laboratory. The NIH has developed a guide to aid proper laboratory commissioning.[29] A similar process occurs in other agencies. The policies and directives established by other agencies generally require comportment with the biosafety guidelines established by HHS through the BMBL.[30]

Agencies generally contract out the process of commissioning a newly constructed or existing laboratory at a given biosafety level. Some agencies have guidelines for this process. For example, NIH suggests that commissioning or certifying occur via companies not involved in the design or construction of a facility. The NIH *Model Commissioning Guide* also states:

> Individuals should be highly specialized in the types of facilities and systems being installed. Due to the degree of technical oversight which is expected, individuals should be licensed Professional Engineers (or as applicable for specialized systems/facilities) with extensive experience in the design, optimization, remediation, and acceptance testing of applicable systems as well as training and building manual preparation.[31]

Grant and Contract Language

Laboratory facilities that receive federal funding, even if not owned or operated by the federal government, generally agree, as a funding condition, to adhere to the biosafety guidelines set forth in the BMBL. If these laboratories are found not to be in compliance with the guidelines, existing federal funding could be withdrawn and future federal funding withheld. For example, the NIH *Grants Policy Statement* states:

> Grantee organizations are not required to submit documented assurance of their compliance with or implementation of these regulations and guidelines. However, if requested by the awarding office, grantees should be able to provide evidence that applicable Federal, State, and local health and safety standards have been considered and have been put into practice.[32]

Similar language may be inserted into contracts and other award mechanisms.

Institutional Biosafety Committees

Federal research grants dealing with recombinant DNA research generally require that the grantee create an institutional biosafety committee (IBC) to oversee the grantee's compliance with federal guidelines addressing such research. One role of the IBC is to determine the appropriate biosafety level for novel recombinant DNA research. An IBC thus has the authority to require that recombinant DNA research activities be performed with specific biosafety precautions or not be performed at that institution. While this authority does not currently extend past recombinant DNA research, some have proposed using the IBCs as a mechanism to oversee other research with biosafety concerns.[33] The BMBL suggests that IBCs help scientists to determine the appropriate level of protection for specific proposed experiments.[34] Some nongovernmental experts, however, have objected that the IBC mechanism does not provide sufficient oversight or rigorous review for such responsibilities.[35]

Select Agent Program

The Select Agent Program regulations incorporate the BMBL guidelines by reference.[36] This incorporation makes consideration of the BMBL a mandatory component of compliance with the Select Agent Program. Entities possessing select agents must develop the security plan discussed previously in "Biosecurity and the Select Agent Program" and a biosafety plan which takes into account the BMBL guidelines. Both of these plans may be reviewed by federal officials as part of the oversight process.

PROLIFERATION OF LABORATORIES

The number of BSL-3 and BSL-4 laboratories is generally believed to have increased in recent years. While the number of BSL-3 laboratories is not known, the total amount of planned or existent BSL-4 space in the United States has increased by an estimated twelve-fold since 2004.[37] This expansion of high-containment laboratory space resulted from federal construction, increased federal funding of research and development activities requiring high-containment laboratories, and a greater focus on public health and diagnostic laboratory capacity.

Federal Laboratories

The federal government is constructing or funding the construction of numerous high-containment laboratories. This section provides examples of federal investment in biocontainment laboratories in the four departments principally sponsoring this expansion: the Departments of Defense, Homeland Security, Health and Human Services, and Agriculture. Some of the facilities are federally owned and operated, some are federally owned but privately operated, and some are privately owned and operated. Thus, the federal investment in high-containment laboratories is creating both government and private infrastructure.

Department of Defense

The Department of Defense (DOD) has maintained a high-containment laboratory infrastructure for many years. These laboratories arose during World War II and have been modernized as research activities have evolved.[38] These laboratories performed both offensive and defensive research, i.e. they developed biological weapons and the means to protect troops against the use of biological weapons. In 1969, President Nixon discontinued the U.S. offensive biological weapons program.[39] All DOD biological laboratories perform solely defensive work, mainly focused on developing countermeasures to protect the warfighter or on preventive clinical research. The United States Army Medical Research Institute for Infectious Diseases (USAMRIID) in Frederick, MD, is the sole location of BSL-4 laboratory space in DOD, but BSL3 laboratory space is found at other sites.[40]

The DOD is engaged in a $683 million expansion of its facilities at USAMRIID that will increase its BSL-3 and BSL-4 laboratory space.[41] A number of other DOD laboratories have also either increased their BSL-3 capacity or increased their degree of containment since 2001.[42] The DOD maintains its BSL-3 and BSL-4 laboratories for a number of reasons, including basic and applied research into infectious disease, evaluation of vaccines and other biological countermeasures, and testing for naturally occurring disease outbreaks under the Global Emerging Infections Surveillance and Response System.[43]

Department of Homeland Security

In the aftermath of the anthrax mailings, the federal government determined that it needed a specialized facility to analyze pathogens for specific characteristics that could help identify perpetrators of a biological attack. President Bush assigned this bioforensic responsibility in 2004 to the Department of Homeland Security (DHS).[44] To meet this responsibility, the DHS established the National Bioforensics Analysis Center as part of the National Biodefense Analysis and Countermeasures Center.[45] These laboratories are currently located in interim facilities at USAMRIID. Permanent facilities are under construction, also in Frederick, MD, including a BSL-4 laboratory and several BSL-3 laboratories to accommodate both bioforensic and biological threat assessment activities. These facilities are expected to become fully operational in 2010.[46]

The Homeland Security Act of 2002 (P.L. 107-296) transferred the Plum Island Animal Disease Center (PIADC), which contains BSL-3 facilities, from USDA to DHS. The DHS has determined that PIADC has reached the end of its usable lifespan. While maintaining the existing PIADC facility, the DHS is in the process of establishing the National Bio- and Agro-Defense Facility (NBAF), a BSL-4 laboratory, to replace PIADC.[47] This new high-containment facility will have the main goal of performing research on pathogens affecting animals and developing countermeasures against these pathogens. When complete, it will take over the research projects performed at PIADC and expand a currently limited research capability for animal pathogens requiring the highest level of biocontainment. The DHS announced its decision to site the NBAF in Kansas.[48]

Department of Health and Human Services

The Department of Health and Human Services (HHS) currently maintains BSL-4 laboratories at the National Institutes of Health (NIH) in Bethesda, MD, and at the Centers for Disease Control and Prevention in Atlanta, GA.[49] Additionally it has invested in construction of two BSL-4 National Biocontainment Laboratories (NBLs) and thirteen BSL-3 Regional Biocontainment Laboratories (RBLs).

The NBLs and RBLs are to serve as a national resource for conducting clinical and laboratory research and testing on pathogens in support of the NIH National Institute of Allergy and Infectious Diseases' biodefense research agenda. Additionally, both are expected to be available to assist public health efforts during a bioterrorism or emerging infectious disease emergency.[50] The two NBLs are being built in Boston, MA, and Galveston, TX.[51] The RBLs are geographically dispersed throughout the United States.[52] The NBLs and RBLs are being constructed through a grant-making process and will be privately owned and operated.[53]

The HHS has also developed the Laboratory Response Network (LRN), a network of laboratories that engage in public health activities. The LRN is charged with maintaining an integrated network of laboratories that can respond to bioterrorism, chemical terrorism and other public health emergencies.[54] The LRN includes federal and state public health facilities, medical institutions, and others. These laboratories are primarily engaged in diagnostic and public health testing of samples, especially in an emergency situation where additional capacity for such testing is needed. Many of these laboratories possess BSL-3 capabilities.

Department of Agriculture

The Department of Agriculture (USDA) will conduct research in the new NBAF being built by DHS. The USDA is also constructing new BSL-3 laboratories. These include laboratories in Ames, IA, consolidating animal health research, diagnosis, and product evaluation in a single facility, and at the National Wildlife Research Center in Fort Collins, CO.

University Laboratories

Academic institutions are investing in high-containment laboratory facilities for a variety of reasons. As federal funding for biodefense research and development has increased, universities have responded by building more high-containment laboratory space. Such laboratory space is seen as increasing their faculty's ability to compete for federal funding and helping the university recruit new faculty. The University of Minnesota, Purdue University, and Ohio State University are examples that have constructed or plan to construct additional BSL-3 laboratory space.[55]

Industry and Non-Profit Laboratories

Private sector companies and non-profit institutions also maintain high-containment laboratory facilities. Pharmaceutical and other companies may need such facilities for medical testing and evaluation, animal efficacy studies, and other product development purposes. Some large companies have determined that the costs of an in-house high-containment laboratory are justified by their product development and manufacturing goals.[56] Others have chosen to contract with outside firms to obtain this capability. These contract firms include Lovelace Respiratory Research Institute, Battelle Memorial Institute, Southern Research Institute, and others.[57]

Issues for Congress

The expansion of high-containment laboratories has raised several issues for policymakers. These include how to determine the appropriate capacity to meet national needs, whether oversight of facilities or personnel need to be increased, balancing the desire for additional laboratory capacity with its accompanying increase in risk, possible international ramifications, and local concerns.

How Much Capacity Is Enough?

After the 2001 anthrax mailings, it became apparent to some policymakers that the existing high-containment laboratory infrastructure was insufficient to meet the needs of that crisis. Congress and the administration decided that additional high-containment laboratory space should be constructed. The high capital and maintenance costs of these facilities were deemed to require federal investment and support. At the same time, these high costs make it important to avoid over-building capacity and to fully utilize existing resources.

Decisions to build and support high-containment laboratories were made in multiple agencies and in multiple budget cycles. Agencies considered their individual requirements, without any robust national needs assessment or coherent, coordinated expansion or utilization plan.[58] As a consequence, policymakers have expressed concern that the new high-containment laboratory capacity may now exceed the national need or the amount that can be operated safely.[59]

A lack of information on existing federal and non-federal high-containment laboratory capacity is hindering more coordinated planning. The National Institute of Allergy and Infectious Diseases identified 277 domestic BSL-3 laboratories, but its survey suffered from a low response rate and other methodological shortcomings.[60] In compliance with the Project BioShield Act of 2004 (P.L. 108-276), DHS and HHS estimated that there are 630 BSL-3 and BSL-4 laboratories.[61] This estimate has also been criticized by nongovernmental experts.[62] The DHS and HHS estimate drew heavily on the number of facilities registered to work with select agents. However, a BSL-3 laboratory need not necessarily work with select agents and therefore may not be required to hold a registration certificate. Conversely, entities registered to work with select agents may or may not use high-containment laboratories (see **Table 3**). Other, nongovernmental estimates, though incomplete, have identified large increases in high-containment laboratory capability.[63] The Government Accountability Office was unable to definitively determine the number of BSL-3 laboratories, but did document an increase in laboratories.[64]

Determining whether ongoing and planned construction has surpassed future national needs likely requires a government-wide assessment with the participation of many federal agencies. Previous efforts by a single agency to determine the number of extant laboratories have failed.[65] Congress may consider whether such a survey should be performed at the agency level or through a higher or external coordinating body.

The continued construction of high-containment facilities raises questions about capacity utilization. Fully utilizing additional capacity will likely require increased funding for research, operations, and management. This increase in funding may be particularly difficult for those agencies that are undergoing a large increase in capacity. For example, an agency increasing its high containment laboratory capacity four-fold will have concomitant increases in operations and maintenance costs for the larger facility regardless of whether the agency fully uses the new capacity. Individual agencies may feel pressure to request larger research budgets to justify their increased operations and maintenance costs. Interagency federal planning efforts may help alleviate overcapacity costs, but would require agreement among agencies regarding prioritization and shared use of different agency facilities.

Planning efforts arising from individual agency initiatives may have difficulty influencing synergies or redundancies between agencies. For example, DHS, through the Under Secretary for Science and Technology, has attempted to coordinate homeland security research and development, but this activity was not prescriptive and did not address use of other agency laboratory assets.[66] Conversely, higher-level, multi-agency strategies may not specify the utilization of existing or future facilities and laboratory space on an agency basis.[67] Congress may wish to consider whether multi-agency coordination of high-containment laboratory use is best performed through central coordination or at the agency level.

A lack of information on existing federal and non-federal high-containment laboratory capacity is hindering more coordinated planning. The National Institute of Allergy and Infectious Diseases identified 277 domestic BSL-3 laboratories, but its survey suffered from a low response rate and other methodological shortcomings.[60] In compliance with the Project BioShield Act of 2004 (P.L. 108-276), DHS and HHS estimated that there are 630 BSL-3 and BSL-4 laboratories.[61] This estimate has also been criticized by nongovernmental experts.[62] The DHS and HHS estimate drew heavily on the number of facilities registered to work with select agents. However, a BSL-3 laboratory need not necessarily work with select agents and therefore may not be required to hold a registration certificate. Conversely, entities registered to work with select agents may or may not use high-containment laboratories (see Table 3). Other, nongovernmental estimates, though incomplete, have identified large increases in high-containment laboratory capability.[63] The Government Accountability Office was unable to definitively determine the number of BSL-3 laboratories, but did document an increase in laboratories.[64]

Determining whether ongoing and planned construction has surpassed future national needs likely requires a government-wide assessment with the participation of many federal agencies. Previous efforts by a single agency to determine the number of extant laboratories have failed.[65] Congress may consider whether such a survey should be performed at the agency level or through a higher or external coordinating body.

The continued construction of high-containment facilities raises questions about capacity utilization. Fully utilizing additional capacity will likely require increased funding for research, operations, and management. This increase in funding may be particularly difficult for those agencies that are undergoing a large increase in capacity. For example, an agency increasing its high containment laboratory capacity four-fold will have concomitant increases in operations and maintenance costs for the larger facility regardless of whether the agency fully uses the new capacity. Individual agencies may feel pressure to request larger research budgets to justify their increased operations and maintenance costs. Interagency federal planning efforts may help alleviate overcapacity costs, but would require agreement among agencies regarding prioritization and shared use of different agency facilities.

Planning efforts arising from individual agency initiatives may have difficulty influencing synergies or redundancies between agencies. For example, DHS, through the Under Secretary for Science and Technology, has attempted to coordinate homeland security research and development, but this activity was not prescriptive and did not address use of other agency laboratory assets.[66] Conversely, higher-level, multi-agency strategies may not specify the utilization of existing or future facilities and laboratory space on an agency basis.[67] Congress may wish to consider whether multi-agency coordination of high-containment laboratory use is best performed through central coordination or at the agency level.

Sufficiency of Current Oversight and Enforcement

The growth of BSL-3 and BSL-4 laboratories has raised concerns about the potential for pathogen release into local communities, as well as biological weapon proliferation, either through the transfer of pathogens or the transfer of technical knowledge through training and employment of foreign scientists in such venues.[68] Events, such as laboratory infections with select agents,[69] improper shipping of select agents,[70] and performance of research at inadequate biosafety levels[71] have led policymakers to reexamine the current oversight framework. Consensus has not been reached regarding the best approaches to overseeing these laboratories. The current, mainly self-regulatory approach is seen by some analysts as not stringent enough and in need of additional federal oversight. Some experts have suggested expanding regulations to require mandatory biosafety training or broadening biosecurity measures.[72] Others feel that the broad application of the results and processes of high-containment laboratory work may lend itself better to an enhanced self-regulatory approach, focused on greater worker engagement and responsibility. These experts suggest that training in best practices be increased and that scientists develop more robust self-policing.[73]

Threat Associated with Increased Number of Facilities

Increased high-containment laboratory capacity is a two-edged sword. Expansion allows for a greater diversity of biodefense research, more efficient public health sample testing, and more research discoveries. However, the increase in laboratories also increases the potential for theft or accidental release of dangerous pathogens and transfer of technical knowledge to persons wishing to do harm.

Legislation passed by previous Congresses addressed concern about the theft of dangerous pathogens by augmenting the Select Agent Program to enhance the physical security of stored pathogens, as discussed in "Biosecurity and the Select Agent Program". As the number of high-containment facilities has expanded, more facilities have registered to possess select agents. Investigations by GAO and agency inspectors general have revealed security weaknesses at several facilities participating in the Select Agent Program.[74] Additional regulatory focus on Select Agent Program compliance might minimize the likelihood of these events, though growing laboratory capacity may strain agency resources for these endeavors.

Whether public or private sector, high-containment laboratories are planned and designed to minimize the possibility of accidents from human error or mechanical failure. Despite such planning, accidents do occur and can overwhelm the safety controls and barriers designed to mitigate their consequences.[75] The GAO has testified that each high-containment facility has an inherent associated risk. [76] As additional facilities are constructed, the total risk increases.

The magnitude of this incremental risk is difficult to measure and may vary depending on each laboratory's compliance with best practices and regulation. Factors that might contribute to the risk include the rate of accidental or other releases, the rate of subsequent illness from any releases, and the total number of laboratories. Because this information is not generally available, policymakers may perform an approximate cost/benefit analysis relying on estimates of the potential benefits provided by research in high-containment laboratories and the potential costs resulting from an infectious release from a high-containment laboratory. For example, breakthroughs in biodefense research, such as vaccines against pathogens requiring high-containment, might be considered as a benefit while the loss of infected animals or infection of laboratory workers might be used as an example of a cost. Subtle and institutional effects, which may be hard to measure, such as the development of a national sampling laboratory network and the proliferation of high-containment laboratory technique, may be overlooked by federal policymakers. As a result, policy may be driven by response to high profile incidents rather than by judging the actual risk, or the full costs and benefits, of the increased number of facilities.

Another concern is personnel surety or the "insider threat." The expansion of high-containment laboratories requires additional trained technicians, scientists, and other employees to utilize this capacity. As more people receive training in high-containment technique, the risk that some of them have malicious intent increases. Some experts have asserted that such skills are not necessary to carry out a bioterrorist attack, but others believe that they are key.[77] The DOJ asserts that the perpetrator of the 2001 anthrax mailings was a government scientist expert in high-containment technique.[78]

International Issues

The increase in high-containment capacity may have international ramifications. Other countries may perceive the increase in high-containment laboratory capacity as beyond that necessary for biodefense and health research. Such a perception might lead some governments to conclude that the United States is constructing this excess capacity for offensive biological weapons development.[79] Such a perception would likely damage U.S. efforts to persuade other countries to adhere to the Biological Weapons Convention.

Expansion of the U.S. infrastructure may also encourage construction of similar facilities in foreign countries in an effort to match U.S. efforts.[80] Such international laboratory proliferation may lead to greater international availability of high-containment technique and information and a greater risk that this training will be used for malevolent purposes.

Differences in biosafety regulation and oversight among countries may influence the business decisions of pharmaceutical and biotechnology companies. If multinational corporations consider the United States to have an unfavorable regulatory environment, they may tend to locate their research and production facilities in other countries. Apart from the possible economic effects of this decision, the location of pharmaceutical and biotechnology companies could also affect national capabilities to respond to bioterrorism or natural disease outbreaks. Countries that rely on foreign production of disease treatments might be at risk of reduced or restricted access to countermeasures during an international disease outbreak or bioterrorist attack. Governments may choose to hoard domestically produced countermeasures to treat their residents rather than allow the countermeasures to leave the country.[81]

Differences in biosecurity regulation between countries have led some U.S. scientists to curb their international collaborations.[82] Because many scientists believe that international collaborations are a key component of the scientific process, this development may have a chilling effect. Perceived barriers to international collaboration between U.S. researchers and those in other countries may lead industry to locate or invest in research in countries where such barriers are perceived to be lower. The extent to which biosecurity regulations have hampered international collaboration, or research in general, remains ambiguous. Additionally, the presence of U.S. biosecurity standards may encourage international partners to develop similar or parallel policies harmonized with U.S. efforts.[83]

Local Concerns

The construction of high-containment laboratories in the United States has been met with varying degrees of local resistance. Some groups have strongly supported the building of new laboratories in their communities, citing the importance of the mission and the likelihood of local economic benefits.[84] Proponents state that high-containment laboratories will generate high-paying jobs and require ancillary support from community businesses, providing a boost to the local economy.

Other groups have strongly opposed proposed new laboratories for public health and quality of life reasons, citing possible release of pathogens and increases in security and traffic.[85] These groups are dubious of economic benefit claims by proponents, questioning whether the high-paying jobs will be available to the existing local workforce rather than individuals recruited from elsewhere.

POLICY OPTIONS

In addressing these issues policymakers have several options. Interested congressional policymakers could deem current efforts sufficient and no further action is warranted. Several expert panels are examining these issues currently and Congress could defer additional action until their reports are complete and their recommendations heard. Alternatively, Congress could decide to step up current oversight of facilities, pathogens, or personnel, or to take other actions.

Status Quo

Oversight efforts attempt to balance security controls with research productivity, weighing the potential for an adverse event against that of scientific progress. One scientist summarized the potential trade-offs between additional oversight and scientific progress: "While, in principle, the scientific community supports measures that would reduce the threat of terrorists acquiring deadly pathogens, there also is strong opposition to measures that make it more difficult to perform research."[86] Policymakers may determine that current oversight efforts regarding high-containment laboratories are sufficient and that additional measures may cause undue burdens on scientific progress. Some portion of the scientific community is likely to support relying on extant oversight mechanisms and taking no additional action.

Supporters of the status quo might argue that securing high-containment laboratories in the United States provides less increased security than focusing federal efforts on identifying potential bioterrorists and disrupting individual bioterrorist activities.[87] Additional resources dedicated to developing intelligence on potential bioterror groups, their organizational needs, and social aspects might allow for disruption of their activities at a lower total cost than the implementation of an overarching regulatory framework that affects, by and large, domestic scientists engaged in beneficial activities.

Await Recommendations

Even if Congress concludes that current oversight efforts are insufficient, concerned policymakers may choose to defer action until they can obtain a fuller understanding of policy options and their impacts.[88] Since 2005, the Bush Administration created at least three expert groups to examine aspects of biosafety and biosecurity issues: the Working Group on Strengthening the Biosecurity of the United States, the Trans-Federal Task Force on Optimizing Biosafety and Biocontainment Oversight, and the National Science Advisory Board for Biosecurity. Additionally, GAO is continuing to investigate the proliferation of high-containment laboratories. Congress could choose to wait until some or all of these groups have completed their efforts and made their recommendations. To make its intent clear, Congress could explicitly endorse the groups' activities while awaiting their results. Additionally some policymakers may deem these efforts as insufficient and may require the performance of additional studies.

President Bush created the Working Group on Strengthening the Biosecurity of the United States on January 9, 2009, through Executive Order.[89] This working group, co-chaired by the Secretaries of Defense and Health and Human Services, consists of the Secretaries of State, Agriculture, Commerce, Transportation, Energy, and Homeland Security; the Attorney General; the Administrator of the Environmental Protection Agency; the Director of National Intelligence; and the Director of the National Science Foundation.[90] It is charged with reviewing and evaluating existing laws, regulations, guidances, and practices of physical, facility, and personnel security and assurance. Its recommendations to the President regarding existing and new biosafety and biosecurity laws, regulations, guidances, and practices are due in July 2009.

The HHS announced the formation of the Trans-Federal Task Force on Optimizing Biosafety and Biocontainment Oversight during congressional hearings in October 2007.[91] This group, established in December 2007,[92] is to develop an options paper that:

- addresses the current framework for local and federal biosafety and biocontainment oversight of research at high-containment laboratories,
- addresses potential gaps in biosafety and biocontainment oversight of high-containment laboratories, and
- provides recommendations for closing these gaps.[93]

Co-chaired by HHS and USDA, the task force includes staff from the Environmental Protection Agency, the National Science Foundation, and the Departments of Commerce, Defense, Energy, Homeland Security, Labor, Transportation, and Veterans Affairs. The task force's recommendations will be presented to HHS and USDA leadership.

The National Science Advisory Board for Biosecurity includes non-governmental voting members and non-voting members from 15 federal agencies and departments. The Board is developing policies relating to dual-use research (i.e., beneficial research that might be turned to malignant uses). Among other endeavors, the Board is trying to develop an effective framework for oversight of such research, a code of conduct for researchers, and effective biosafety training programs for researchers.[94] The Board has multiple working groups producing analysis and publishes recommendations on an ongoing basis.[95]

The GAO is expected to release a report in 2009 that addresses high-containment laboratory issues and builds on prior testimony. [96] This chapter may have additional recommendations for legislative branch and executive branch action.

By waiting for the recommendations of such studies, Congress may be able to use the results of these efforts rather than duplicating them and possibly undercutting them. Additionally, these efforts may generate new policy ideas or options not currently under consideration.

Enhance Oversight of Facilities

Alternatively, Congress may decide that high-containment laboratories issues are too pressing to wait for the results of these groups and instead address this issue without waiting for executive branch or GAO reports. The following are several policy options that may facilitate this approach.

Survey of Facilities and Use

Previous efforts to determine the number and capacity of such laboratories have failed to produce a robust and reliable determination of the number of high-containment laboratories. Multiple sources, including non-governmental organizations and congressionally established commissions, have called for a comprehensive inventory of existing capacity.[97]

While compiling such a comprehensive inventory may appear straightforward, it likely would not be an easy or simple task. Non-federal laboratories reside in a variety of locations performing a variety of work. Non-federal high-containment laboratories are located in academic research institutions, public health agencies, and industrial research and quality control facilities. A voluntary self-reporting mechanism may be insufficient to identify existing facilities and get information about their capacity. Also, a survey would be a snapshot of the actual high-containment capacity but would not necessarily capture future construction or expansion. Concerns about industrial competitiveness and reporting burden may act as substantial barriers for obtaining a complete inventory. [98] Efforts to incentivize self-reporting might increase participation depending on the incentive relative to the reporting barrier.

Alternatively, congressional policymakers could mandate that all high-containment laboratories, both public and private, register with the federal government. If laboratories face a sufficient penalty for not registering, this approach could produce a comprehensive list of high-containment laboratories. Such a registry could potentially provide the government with information regarding the total available laboratory capacity, its current use, its geographic distribution, and availability. Facility registration was a component of early implementation of the Select Agent Program, prior to amendments requiring certification of facilities and personnel. Registration might be required even if no further security measures were mandated.

Another option, given the difficulty of determining the full scope of high-containment laboratories, is to limit such a survey to federal facilities. Since federal facilities are within the scope of executive decree, all agencies could be required to report their high-containment facilities and capacity to Congress through the Executive Branch. If the existing federal capacity was well known by policymakers, the government might more efficiently plan for use and possible expansion of this capacity. A concomitant development of a government-wide needs assessment may also help the government to efficiently apportion resources. By comparing the needs assessment with existing and planned federal capacity, it may be possible for the government to determine if federal high-containment capacity is greater than the need. Overexpansion has both resource allocation costs and potentially increased security concerns.

In considering whether to attempt to survey the number of high-containment laboratories, Congress may wish to weigh the potential economic costs of doing so under both a self-reporting framework and a mandatory reporting framework, as well as the quality and utility of the information so acquired. Additionally, the benefits of such a survey may be maximized through comparison with a needs assessment, which is currently unavailable across federal agencies. Concerned Members of Congress might require the submission of a needs assessment from specific agencies or from multiple agencies in addition to capacity information.

Moratorium on Construction

Some policymakers have called for a moratorium on new federally funded construction of high-containment laboratories and a decrease in the proposed final number of laboratories.[99] In weighing the effects of a moratorium, policymakers would likely have to address the benefit of constructing the right amount of laboratory capacity, rather than too much, versus the potential costs to research and other priorities posed by such a moratorium.

In the absence of a government-wide capacity needs assessment, building additional laboratories increases the probability that capacity will be over-built or redundant operations established among agencies. The establishment and operation of new high-containment laboratories is likely to reach a level of diminishing returns as the capacity approaches the level of need. Incremental increases in high-containment laboratory capacity may yield greater return than the construction of large amounts of new laboratory space, suggesting that existing laboratory space could be expanded rather than new laboratory space constructed. Some agencies have countered such criticism by stating that they have assessed interagency needs on a case by case basis and that existing infrastructure is dated and not a candidate for expansion. For example, DHS states it has performed such an assessment as part of its development for the National Bio- and Agro-Defense Facility.[100] However, in the absence of a government-wide assessment, it is possible that agency- specific needs assessments will not identify synergies or existing available infrastructure in other agencies.

Enacting a construction moratorium might have ramifications for the federal government's ability to meet previously established national goals. For example, delays in construction of laboratory facilities may result in delays in performing planned research activities that require those facilities. Previously identified national priorities, such as the establishment of a permanent bioforensics capability, might be delayed or hampered.[101] Additionally, advisory panels and other groups have made repeated calls for expansion of specific high-containment laboratory facilities.[102] A moratorium on construction might impede the development of capabilities deemed necessary by such panels.

Require High-Containment Laboratory Certification

Some analysts have suggested that the federal government license and certify all high-containment laboratories.[103] Currently, only laboratories that handle select agents are required to register with the federal government and receive certification that they comply with specified security regulations. Among the possible approaches are directly regulating high-containment laboratories and expanding the scope of the Select Agent Program. An analysis of these two options follows.

Directly Regulate High-Containment Laboratories

One approach to licensing might be applying the Select Agent Program requirements to all high-containment laboratories, regardless of the pathogens used at the laboratory. Such an approach might provide a uniform set of expectations regarding the security level at all high-containment laboratories.

Alternatively, licensing could be another, separately administered program with its own requirements. Establishing a separately administered program might allow for flexibility in licensing requirements, differentiating, for example, between high-containment laboratories that possess select agents and those that do not. Laboratory licensing and certification mandates might pose a series of challenges, both in implementation and in acceptance from the scientific community. Licensing and certifying high-containment laboratories may not address the full universe of laboratories of concern. For example, select agents may be handled outside of high-containment laboratories under certain circumstances, indicating that certain manipulation of these pathogens could occur outside of a unified regulatory framework that covers only high-containment facilities. A more comprehensive approach would involve overseeing all laboratories capable of BSL-2 or higher containment. This approach would be certain to capture all locations of sufficient biosecurity for the use of any select agent but would have several disadvantages. Primary among them is the greatly increased number of facilities that would be overseen by this approach. Most facilities handling a human-infective pathogen would qualify under this approach, including and impacting public health, diagnostic, hospital, industrial, and academic laboratories.

Expand the Select Agent List

A different approach, focusing on the pathogens of concern rather than the facilities capable of handling them, might involve expanding the select agent list. As the Select Agent Program uses a list of pathogens to identify regulated entities, some experts have argued that some dangerous pathogens are not captured by this program. Some potentially dangerous diseases not covered by the Select Agent regulations include severe acute respiratory syndrome (SARS), dengue fever, western equine encephalitis, and yellow fever. Some of the pathogens that cause these diseases have been considered as biological weapons.[104] Expanding the number of regulated pathogens would mean that more high-containment facilities would become regulated through the Select Agent Program.

Some scientists have expressed concern about how the select agent list is constructed and may resist an expansion of the list. They argue that the current definition of a select agent, which relies on taxonomic definition of pathogens, may not be specific enough to accurately differentiate pathogens of concern from similar pathogens not of concern and thus may be unnecessarily hindering research.[105] The NIH has requested the National Academies to determine the scientific advances necessary to identify select agents based on other features and properties beyond taxonomy.[106]

By focusing on the Select Agent list, regulatory impacts would affect only those entities using pathogens of concern. Other high-containment facilities would not be affected. Such an approach would potentially limit disruption of scientific development and public health activities, as well as allow the government to focus its regulatory efforts on a subset of the high-containment laboratory universe deemed to pose the greatest risk.

Enhance Oversight of Laboratory Personnel

Current Select Agent Program personnel background checks could be expanded to all personnel using BSL-3 and BSL-4 facilities. Select Agent Program personnel background checks have been identified by the DOJ Inspector General as sharing some of the requirements of other security regimes, such as the background investigation accompanying access to classified information or required for positions of public trust.[107]

Alternatively, a different level of background screening, perhaps less rigorous than used for possession of select agents, might be required for access to high-containment facilities that do not possess select agents. The scope of such background screening is likely to depend on policymakers' evaluation of the potential risks involved with gaining access to high-containment facilities.

Efforts to enhance personnel oversight at high-containment facilities may pose a series of implementation challenges. Primary may be the potential resistance to government documentation and certification of laboratory researchers. Some scientists may assert that they would prefer to exit research fields requiring high-containment laboratories rather than obtain the required government clearance. The extent of this possible resistance is unclear.[108]

Making such personnel screening effective may be difficult. The Department of Defense maintains personnel oversight programs more extensive than required for access to select agents.[109] However, the DOJ has asserted that, despite this increased oversight, a DOD scientist perpetrated the 2001 anthrax mailings.[110]

Standardize Training

The Select Agent Program requires worker training prior to gaining access to select agents as well as annual retraining. Rather than prescribing specific training requirements, the regulations state, "The training must address the particular needs of the individual, the work they will do, and the risks posed by the select agents or toxins."[111] While this allows for flexibility, it also risks allowing variation in quality. The development of training standards could help assure that each trainee receives certain core competencies.

Some experts have suggested that, although there are some exemplary biosafety training programs, these efforts are insufficient to meet current demand and should be expanded.[112] This type of training could be structured in many different ways, ranging from voluntary, local training to mandatory, federally centralized certification of competency. The effects of such training may be difficult to assess.

Expanded biosafety training may reduce the number of laboratory acquired infections, but the rate of these infections is reportedly lower than that for other injuries. Additionally, purely biosafety training may not address biosecurity weaknesses or vulnerabilities. Enhancement of training to include considerations related to the potential dual-use nature of pathogens handled in high-containment laboratories might address concerns related to biosecurity and provide a more uniform understanding of biosecurity concerns among the high-containment laboratory worker community.

Improve Reporting of Lessons Learned

When accidents or other unexpected events occur in high-containment laboratories, lessons may be learned from the success or failure in addressing the particular incident. Reporting mechanisms, both voluntary and mandatory, have been suggested and implemented in other areas to help disseminate information about accidents or near accidents, to help the community learn from the experiences of others.[113] In the case of events involving select agents, the potential for criminal or other penalties to occur may pose a disincentive to reporting these events widely among the research community.[114] Researchers or research institutions might also not report because of perceived loss of stature, embarrassment, or other professional repercussions. This lack of sharing may inhibit the improvement of practices and procedures that might prevent such events from happening in the future.

Some experts have suggested enhancing the current reporting system to establish a database of accidents and corresponding remediation, such as changes in technology or practice. Such a database might provide warning to scientists involved in similar efforts, allowing future accidents to be avoided or minimized. The reporting required under the Select Agent Program could be coupled with such a database. Entries into such a database might be affiliated with the reporting entity or be anonymous after processing by HHS or USDA. Experts have recommended that penalties for reporting accidents to this database be minimized or eliminated so as to encourage such reports.[115] When considering such a mechanism, Congress may have to weigh the potential for negligence to go unpunished when reported and the interaction of such reporting with the existing criminal and other penalties present in the Select Agent Program.

Analysis

When crafting the details of such programs, policymakers may have to assess the optimal balance between increasing oversight and the potential regulatory burden. Approaches that increase federal oversight would be likely met with at least some resistance from scientists and other affected stakeholders. The benefits of such increased oversight might be hard to quantify, expressing themselves mainly as reducing potential opportunities for ill-doers to use existing infrastructure in the pursuit of biological weapons.

Expansion of the Select Agent Program, either through expanding the Select Agent list or by applying the program requirements to more laboratories, may increase barriers to public health response and international collaboration. Critics of the Select Agent Program have stated that the public health response to emerging disease, such as SARS and avian influenza, relies on timely and efficient transfer of materials between high-containment laboratories both domestically and internationally.[116] Increasing the scope of the Select Agent Program, with its requirements for both laboratory and personnel certification, may increase the barriers to successful collaboration, leading to negative impacts on public health response. Similarly, international collaboration on regulated pathogens, both in the public health and the scientific research community, may be impeded because of a lack of comparable security regimes in foreign countries.

The impacts of a licensing and certification regime on the regulated entities might be significant. Scientists not involved in biodefense activities would be affected by the licensure requirement, potentially reducing scientific productivity in both the academic and industrial sectors. Also, public health laboratories and hospitals may accrue costs associated with these regulatory activities.

Moreover, an increase in regulated facilities would require a concomitant increase in federal resources, both to process initial registration and to perform necessary inspections and certification oversight.

One overarching issue with any of these options is the identification of the most appropriate oversight agency. With regard to existing biosafety and biosecurity programs, policymakers determined that USDA and HHS had the necessary technical knowledge and relationship with the scientific community to make those agencies the appropriate regulators. Historically, many of these concerns were more focused on laboratory worker safety and minimizing the risk of pathogen accidental release. As homeland security concerns, such as pathogen theft and the possibility of training someone with malicious intent, have increased, policymakers may decide that another agency, such as DHS, would be more appropriate.

If policymakers augment existing authorities rather than creating new ones, oversight could be performed by the agency currently possessing statutory authority. If policymakers choose to develop new, additional authorities, then a mixture of oversight authority might occur, where one agency regulates for security under the Select Agent Program, and another under a new authority regulates high-containment laboratories. Harmonizing any new requirements with existing requirements, including deciding whether to consolidate all oversight into one agency, may be important to smooth implementation of new oversight responsibilities.

As biosafety is mainly based on voluntary adoption of best practices, limited federal resources have been expended in overseeing such adoption. If the scope of biosafety practices for high-containment laboratories is expanded, then the amount of federal resources necessary, even under a non-mandatory program, would likely increase. Additional funding or staffing may be necessary for those agencies to oversee and meet new mandates.

LEGISLATION IN THE 111TH CONGRESS

The 111th Congress has begun to consider some of these issues. Legislation addressing the Select Agent Program and the questions raised by the expansion of high-containment laboratory capacity have been introduced into both chambers. H.R. 1225 and S. 485 would extend the authorization of the Select Agent Program through 2013 or 2014 respectively and expand the criteria considered by CDC and USDA when determining if pathogens are select agents. The bills would require the National Academy of Sciences to review the Select Agent Program and recommend improvements. The HHS Secretary would be required to develop minimum biosafety and biosecurity training that would be mandatory to gain access to select agents. Additionally, the HHS Secretary would be required to issue guidance on improvements to the current select agent monitoring and inventory procedures.

These bills would also require the HHS Secretary, in consultation with the USDA Secretary, to report to Congress an evaluation of sufficiency of current and planned capacity; how laboratory capacity and lessons learned could be best shared across the biodefense and infectious disease communities; how to improve and streamline guidance on laboratory infrastructure, commissioning, and maintenance; and ways to improve personnel training. Finally, the HHS Secretary is directed, in coordination with the USDA Secretary, to establish the Biological Laboratory Incident Reporting System, through which laboratory personnel could voluntarily report biosafety or biosecurity incidents.

The Laboratory Surge Capacity Preparedness Act (H.R. 1150) would authorize DHS to award competitive grants to Regional Biocontainment Laboratories for their operation and maintenance costs. This bill would also require the DHS Secretary, in consultation with the HHS Secretary, to provide a report to Congress. This chapter would detail activities undertaken to integrate the RBL network and describe whether additional BSL-3 laboratory surge capacity is needed to effectively respond to a biodefense or emerging infectious disease emergency.

LOOKING AHEAD

Regardless of U.S. domestic efforts, biocontainment technologies are widely dispersed around the globe and used by many scientists in many countries. Absent international harmonization, the threat of a high-containment laboratory being the source of a bioterror weapon may be only partially addressed by solely domestic policy changes.

A key challenge for congressional policymakers is to define the goal of enhanced oversight of high-containment laboratories. The choice of goal may affect the relative importance of the issues of concern and thus the choice of policies to address them. For example, focusing on a registry of existing high-containment laboratory capacity may have benefits for planning, coordination, and efficiency of use but provide relatively limited security benefits. Similarly, a rigorous oversight program including facility and personnel licensure, mandatory training, and restricted construction of new facilities may provide security benefits at the cost of regulatory burden, increased federal expenditures, and impeded scientific progress in countermeasure research, bioforensics, and public health. When weighing potential policy options to address these complex policy issues, policymakers may have to reconcile many competing and potentially conflicting national needs.

End Notes

[1] Crystal Franco, "Billions for Biodefense: Federal Agency Biodefense Funding, FY2001-FY2009," *Biosecurity and Bioterrorism: Biodefense Strategy, Practice, and Science*, vol. 6, no. 2 (2008), pp. 131-146. Other groups use different definitions of what constitutes biodefense, but observe increases of similar magnitude. For example, see Alan Pearson, *Federal Funding for Biological Weapons Prevention and Defense, Fiscal Years 2001 to 2009*, The Center for Arms Control and Non-Proliferation, Washington, DC, May 27, 2008, http://www.armscontrolcenter.org/policy/biochem/ articles/fy09 _biodefense _funding/.

[2] Commission on the Prevention of Weapons of Mass Destruction Proliferation and Terrorism, *World at Risk*, December 2008.

[3] The terms can be used differently in other contexts. For example, in agricultural parlance, the term biosecurity may refer to practices to minimize the introduction of plant and animal pathogens to farms.

[4] For more information on the Select Agent Program, see online at http://www.selectagents.gov/.

[5] The Select Agent pathogen lists can be found at 7 C.F.R. 331 (plant pathogens), 9 C.F.R. 121 (animal pathogens), and 42 C.F.R. 73 (human pathogens). An entity is defined as any government agency (federal, state, or local), academic institution, corporation, company, partnership, society, association, firm, sole proprietorship, or other legal entity. An entity is thus not limited to a single facility or to a single laboratory. An entity may possess one or multiple facilities, each facility containing one or multiple laboratories.

[6] Dr. Richard Besser, Director of the Coordinating Office for Terrorism Preparedness and Emergency Response, Centers for Disease Control and Prevention, Testimony before the House Committee on Energy and Commerce, Subcommittee on Oversight and Investigations, October 4, 2007.

[7] Centers for Disease Control and Prevention, personal communication, February 9, 2009; and Department of Agriculture, personal communication, February 4, 2009.

[8] For more information on the Laboratory Response Network, see online at http://www.bt.cdc.gov/lrn/.

[9] Centers for Disease Control and Prevention, Office of Inspector General, Department of Health and Human Services, "Possession, Use, and Transfer of Select Agents and Toxins," 70 *Federal Register* 13294-13325, March 18, 2005.

[10] Department of Energy, Office of Inspector General, *Coordination of Biological Select Agent Activities at Department of Energy Facilities*, DOE/IG-0695, July 2005.

[11] Department of Justice, Office of Inspector General, *Inspection of the FBI's Security Risk Assessment Program for Individuals Requesting Access to Biological Agents and Toxins*, I-2005-003, March 2005.

[12] Department of Health and Human Services, Office of Inspector General, *Summary Report on Universities' Compliance with Select Agent Regulations*, A-04-05-02006, June 2006; and Department of Health and Human Services, Office of Inspector General, *Summary Report on Select Agent Security at Universities*, A-04-04-02000, March 2004.

[13] Department of Health and Human Services, Office of Inspector General, *Summary Report on State, Local, Private, and Commercial Laboratories' Compliance with Select Agent Regulations*, A-04-06-01033, January 2008.

[14] Department of Agriculture, Office of Inspector General, *Audit Report Controls Over Biological, Chemical, and Radioactive Materials at Institutions Funded by the U.S. Department of Agriculture—Government-wide Policies Are Needed to Establish Security Standards for Federally-Funded Research at Non-Federal Institutions*, Report No 50099-14-AT, September 2003; Department of Agriculture, Office of Inspector General, *Animal and Plant Health Inspection Service Evaluation of the Implementation of the Select Agent or Toxin Regulations—Phase I*, Report No 33601-2-AT, June 2005; and Department of Agriculture, Office of Inspector General, *Animal and Plant Health Inspection Service Evaluation of the Implementation of the Select Agent or Toxin Regulations—Phase II*, Report No. 33601-3-AT, January 2006.

[15] Emily Ramshaw, "A&M Lab Employee Lacked Clearance in Bioagent Case; Records Detail Security Breaches at University, Other State Schools," *The Dallas Morning News*, July 8, 2007; Emily Ramshaw, "A&M Admits Errors in Bioagent Case; Interim President Denies Wrongdoing, Questions CDC Timing," *The Dallas Morning News*, July 3, 2007; and Emily Ramshaw, "CDC Suspends A&M Research on Infectious Diseases; CDC Suspends Bioagent Work after Exposures Not Reported Promptly," *The Dallas Morning News*, July 2, 2007.

[16] Fines levied through August 16, 2008. See http://oig.hhs.gov/fraud

[17] Department of Health and Human Services, Centers for Disease Control and Prevention and National Institutes of Health, *Biosafety in Biomedical and Microbiological Laboratories (BMBL)*, 5th edition, 2007, http://www.cdc.gov/OD/ ohs/biosfty/bmbl5/bmbl5toc.htm.

[18] Other agencies have expanded on the BSL-1 to BSL-4 nomenclature through additional guidance. For example, USDA has developed a BSL-3Ag nomenclature and guideline to define BSL-3 containment applicable to large animal research. USDA, Agricultural

Research Service, *ARS Facilities Design Standards*, 242.1-M ARS, July 24, 2002, http://www.afm.ars.usda.gov/ppweb/PDF/242-01M.pdf.

[19] *BMBL*, 5th edition, 2007, http://www.cdc.gov/OD/ohs/biosfty/bmbl5/bmbl5toc.htm.

[20] Gigi Kwik Gronvall and Nidhi Bouri, "Biosafety Laboratories," *Biosecurity and Bioterrorism: Biodefense Strategy, Practice, and Science*, Vol. 6, 2008, pp. 299-3 07.

[21] *BMBL*, 5th edition, 2007, Section VI, http://www.cdc.gov/OD/ohs/biosfty/ bmbl5/bmbl5 toc.htm.

[22] *BMBL*, 5th edition, 2007, http://www.cdc.gov/OD/ohs/biosfty/bmbl5/bmbl5toc.htm.

[23] *BMBL*, 5th edition, 2007, Section VI, http://www.cdc.gov/OD/ohs/biosfty/bmbl5/ bmbl5toc.htm.

[24] CRS analysis of *BMBL*, 5th edition, 2007, http://www.cdc.gov/OD/ohs/biosfty/bmbl5 /bmbl5toc.htm.

[25] Certain experiments with pathogens not on the select agent list should be performed at BSL-4 containment. For example, newly discovered pathogens of unknown human infectiousness that are closely related to pathogens normally handled at BSL-4 should also be handled at BSL-4 until it is determined that they may be safely handled under lesser containment.

[26] Department of Health and Human Services, National Institutes of Health, Office of Research Facilities, *Design Policy and Guidelines*, 2003.

[27] Department of Health and Human Services, National Institutes of Health, Division of Public Safety, *NIH Security Design Policy and Guideline*.

[28] See, for example, Department of Health and Human Services, National Institutes of Health, *Biosafety Level 3- Laboratory Certification Requirements*.

[29] Department of Health and Human Services, *The NIH Model Commissioning Guide*, http://des.od.nih.gov/eWeb/ research/farhad2/Commissioning/nih_cx_guide/ComGuideTitle.htm.

[30] See, for example, U.S. Department of Agriculture, Agricultural Research Service, *USDA Security Policies and Procedures for Biosafety Level-3 Facilities*, Department Manual 9610-001, August 30, 2002. See also Department of Homeland Security, "Listing of Applicable Directives for a National Biodefense Analysis and Countermeasures Center Operating Contract, a Federally Funded Research and Development Center," NBACC FFRDC RFP Attachment # 5, May 15, 2006.

[31] Department of Health and Human Services, *The NIH Model Commissioning Guide*, http://des.od.nih.gov/eWeb/ research/farhad2/Commissioning/nih_cx_guide/ComGuideTitle.htm.

[32] Department of Health and Human Services, National Institutes of Health, *NIH Grants Policy Statement*, December 1, 2003, p. 49.

[33] For example, the National Science Advisory Board for Biosecurity (NSABB), an HHS federal advisory board, has recommended empowering the IBCs to review research that has implications for both defensive and offensive use. For more information on the NSABB, see CRS Report RL33342, *Oversight of Dual-Use Biological Research: The National Science Advisory Board for Biosecurity*, by Dana A. Shea.

[34] *BMBL*, 5th edition, 2007, Section II, http://www.cdc.gov/OD/ohs/biosfty/bmbl5/bmbl5toc.htm.

[35] The Sunshine Project, *Mandate for Failure: The State of Institutional Biosafety Committees in an Age of Biological Weapons Research*, October 2004.

[36] 42 C.F.R. 72.13.

[37] Edward Hammond, Director, The Sunshine Project, Testimony before the House Committee on Energy and Commerce, Subcommittee on Oversight and Investigations, October 4, 2007.

[38] For an overview of the establishment and development of Department of Defense laboratories, see David R. Franz, Cheryl D. Parrott, And Ernest T. Takafuji, "The U.S. Biological Warfare and Biological Defense Programs," in *Medical Aspects of Chemical And Biological Warfare*, (Washington, D.C: Office of The Surgeon General, United States Army, 1997).

[39] Executive Office of the President, National Security Decision Memorandum 35, November 25, 1969. The closely related offensive program using biologically derived toxins was ended

three months later. Executive Office of the President, National Security Decision Memorandum 44, February 20, 1970.

[40] Department of Defense, *Chemical and Biological Defense Program Annual Report to Congress*, April 2007, p. 63.

[41] Department of Defense, *FY2009 Defense-Wide Military Construction Justification, Defense Medical Program, TRICARE Management Activity*, p. 29.

[42] Department of Defense, *Chemical and Biological Defense Program Annual Report to Congress*, April 2007.

[43] For overviews of these activities, see Department of Defense, *Chemical and Biological Defense Program Annual Report to Congress*, April 2007, and Department of Defense Global Emerging Infectious Surveillance and Response System at http://www.geis.fhp.osd.mil/.

[44] Executive Office of the President, The White House, *Biodefense for the 21st Century*, Homeland Security Presidential Directive-10, April 28, 2004.

[45] See CRS Report RL32891, *The National Biodefense Analysis and Countermeasures Center: Issues for Congress*, by Dana A. Shea.

[46] J. Patrick Fitch, Laboratory Director, NBACC, and President, Battelle National Biodefense Institute, *Science-Based Biodefense Analysis in an Uncertain World*, Presentation at the Center for Strategic and International Studies, May 12, 2008, http://www.csis.org/media

[47] See CRS Report RL34 160, *The National Bio- and Agro-Defense Facility: Issues for Congress*, by Dana A. Shea, Jim Monke, and Frank Gottron.

[48] Department of Homeland Security, "Science and Technology Directorate; Record of Decision for the National Bio and Agro-Defense Facility Environmental Impact Statement," 74 *Federal Register* 3065-3080, January 16, 2009.

[49] Although built as a BSL-4 laboratory, the Bethesda laboratory operates at the BSL-3 level. See National Institute of Allergy and Infectious Diseases, *The Need for Biosafety Laboratory Facilities*, September 2007, http://www.niaid.nih.gov/factsheets/facilityconstruct_06.htm.

[50] Department of Health and Human Services, *Broad Agency Announcement: Regional Biocontainment Laboratories (RBL) and National Biocontainment Laboratories (NBL)*, BAA-NIH-NIAID-NCRR-DMID-03-36, February 10, 2003; and Department of Health and Human Services, *Request for Applications: Regional Biocontainment Laboratories (RBL) Construction Program*, RFA-AI-04-032, June 29, 2004.

[51] For more information about the NBL being built in Boston, MA, see online at http://www.bu.edu/dbin/neidl/en/. For more information about the NBL being built in Galveston, TX, see online at http://www.utmb.edu/gnl/.

[52] For an overview of the status and locations of the NBLs and the RBLs, see online at http://www3.niaid.nih.gov/LabsAndResources/resources

[53] For an overview of the original proposal for construction of the NBLs and RBLs, see online at http://www.niaid.nih.gov/contract/archive/BAA0336-0.pdf.

[54] For more information on the Laboratory Response Network, see online at http://www.bt.cdc.gov/lrn/.

[55] For information regarding the University of Minnesota's efforts, see online at http://www1.umn.edu/groots/pdf/ MBB.pdf. For information regarding Purdue University's efforts, see online at http://www2.itap.purdue.edu/bot/memberDocuments/ Stated MeetingFiles/animal%20disease%20diagnostic%20laboratory%20_bsl-3_-desc-plan%20res%20_2_1a.pdf. For information regarding the Ohio State University's efforts, see online at http://researchnews.osu.edu/archive/bsl3%20lab.htm, http://researchnews.osu.edu/archive/new%20bsl3.htm, http://medicalcenter.osu.edu/ research/ research tower/, and http://researchnews.osu.edu/archive/bsl3qanda.htm.

[56] See, for example, Dr. Bradley E. Kosiba, Senior Director, Wyeth Global Research Facilities, "Wyeth's New (2006) Containment Facility (BSL3-LS) for the Bio-manufacturing of Bacterial Vaccines," Presentation at the 2007 International Conference on Biocontainment Facilities, March 13, 2007.

[57] See http://www.lrri.org/DPage.aspx?p=42; http://www.battelle.org/solutions/default.aspx?Nav_Area=Solution&Nav_SectionID=4&Nav_CatID=4_ThreatAssessment Characterization; and http://www. southernresearch.org/contractservices/emerging-pathogens

[58] See U.S. Congress, House Committee on Energy and Commerce, Subcommittee on Oversight and Investigations, *Hearing on Germs, Viruses, and Secrets: The Silent Proliferation of Bio-Laboratories in the United States*, 1 10th Cong., 1st sess., October 4, 2007, and U.S. Government Accountability Office, *High-containment Biosafety Laboratories: Preliminary Observations on the Oversight of the Proliferation of BSL-3 and BSL-4 Laboratories in the United States*, GAO-08-108T, October 4, 2007.

[59] See U.S. Congress, House Committee on Energy and Commerce, Subcommittee on Oversight and Investigations, *Hearing on Germs, Viruses, and Secrets: The Silent Proliferation of Bio-Laboratories in the United States*, 1 10th Cong., 1st sess., October 4, 2007.

[60] Constella Health Sciences, *Survey for Determining the Location, Capacity, and Status of Existing and Operating BSL-3 Laboratory Facilities within the United States*, September 9, 2005. The GAO testified that this survey has methodological difficulties. See Sushil Sharma, GAO, Testimony before the House Committee on Energy and Commerce, Subcommittee on Oversight and Investigations, October 4, 2007.

[61] Department of Health and Human Services and the Department of Homeland Security, *Report to Congress Regarding Biocontainment Facilities*, January 2007.

[62] For example, see Alan M. Pearson, Center for Arms Control and Non-Proliferation, Testimony before the House Committee on Energy and Commerce, Subcommittee on Oversight and Investigations, October 4, 2007.

[63] Edward Hammond, Director of The Sunshine Project, Testimony before the House Committee on Energy and Commerce, Subcommittee on Oversight and Investigations, October 4, 2007.

[64] U.S. Government Accountability Office, *High-containment Biosafety Laboratories Preliminary Observations on the Oversight of the Proliferation of BSL-3 and BSL-4 Laboratories in the United States*, GAO-08-108T, October 4, 2007.

[65] See "Survey of Facilities and Use" for further discussion of this point.

[66] Department of Homeland Security, Science and Technology Directorate, *Coordination of Homeland Security Science and Technology*, December 2007 (Revised January 2008).

[67] See, for example, National Science and Technology Council, Committee on Homeland and National Security, Subcommittee on Foreign Animal Disease Threats, *Protecting Against High Consequence Animal Diseases: Research & Development Plan for 2008-2012*, January 22, 2007.

[68] Such concerns have been reportedly identified and raised in the United Kingdom by British security services. Mark Townsend, "Terrorists Try to Infiltrate UK's Top Labs," *The Observer*, November 2, 2008, online at http://www.guardian. For related concerns regarding proliferation of technical knowledge, see Eileen Choffnes, "Biological Weapons: New Labs, More Terror?" *Bulletin of the Atomic Scientists*, Vol. 58, No. 5 (September/October 2002), pp. 28-32. See also, CRS Report RL32891, *The National Biodefense Analysis and Countermeasures Center: Issues for Congress*, by Dana A. Shea.

[69] Emily Ramshaw, "A&M Lab Employee Lacked Clearance in Bioagent Case; Records Detail Security Breaches at University, Other State Schools," *The Dallas Morning News*, July 8, 2007; Emily Ramshaw, "A&M Admits Errors in Bioagent Case; Interim President Denies Wrongdoing, Questions CDC Timing," *The Dallas Morning News*, July 3, 2007; and Emily Ramshaw, "CDC Suspends A&M Research on Infectious Diseases; CDC Suspends Bioagent Work after Exposures Not Reported Promptly," *The Dallas Morning News*, July 2, 2007.

[70] Martin Enserink and Jocelyn Kaiser, "Accidental Anthrax Shipment Spurs Debate Over Safety," *Science*, June 18, 2004, pp. 1726-1727.

[71] "Study of Ebola Virus in U.S. Lab Halted," *Associated Press*, September 20, 2007.

[72] American Society for Microbiology, "ASM Comments on Laboratory Biosafety," *Letter to Chairman Bart Stupak, House Committee on Energy and Commerce, Subcommittee on Oversight and Investigations*, October 3, 2007.

[73] Gigi Kwik Gronvall, Joe Fitzgerald, Allison Chamberlain, et al., "High-Containment Biodefense Research Laboratories: Meeting Report and Center Recommendations," *Biosecurity and Bioterrorism: Biodefense Strategy, Practice, and Science*, vol. 5, no. 1 (2007), pp. 75-85.

[74] Department of Health and Human Services, Office of Inspector General, *Summary Report on State, Local, Private, and Commercial Laboratories' Compliance with Select Agent Regulations*, A-04-06-01033, January 2008; Department of Health and Human Services, Office of Inspector General, *Summary Report on Universities' Compliance with Select Agent Regulations*, A-04-05-02006, June 2006; and Department of Health and Human Services, Office of Inspector General, *Summary Report on Select Agent Security at Universities*, A-04-04-02000, March 2004. See also U.S. Government Accountability Office, *Biosafety Laboratories: Perimeter Security Assessment of the Nation's Five BSL-4 Laboratories*, GAO-08-1092, September 2008; and Commission on the Prevention of Weapons of Mass Destruction Proliferation and Terrorism, *World at Risk*, December 2008.

[75] Department of Homeland Security, *National Bio and Agro-Defense Facility Draft Environmental Impact Statement*, June 2008, pp. 3-364, http://www.dhs.gov/xres/labs/gc_1187734676776.shtm.

[76] U.S. Government Accountability Office, *High-containment Biosafety Laboratories Preliminary Observations on the Oversight of the Proliferation of BSL-3 and BSL-4 Laboratories in the United States*, GAO-08-108T, October 4, 2007.

[77] For a discussion of the diversity of opinion regarding the ease with which biological weapons can be made by non- state groups, see Milton Leitenberg, "An Assessment of the Biological Weapons Threat to the United States," *The Journal of Homeland Security*, January, 2001.

[78] Department of Justice, *Transcript of Amerithrax Investigation Press Conference*, August 6, 2008, http://www.justice. See also Scott Shane, "Portrait Emerges of Anthrax Suspect's Troubled Life," *The New York Times*, January 3, 2009.

[79] For a discussion of this issue, see Milton Leitenberg, *Assessing the Biological Weapons and Bioterrorism Threat*, December 2005, http://www.strategicstudiesinstitute.army.mil/pubs/display.cfm?pubID=639.

[80] Jonathan Tucker, "Biological Threat Assessment: Is the Cure Worse than the Disease?," *Arms Control Today*, Vol. 34, No. 8, October 2004, pp. 13-19.

[81] Michael Osterholm, "Preparing for the Next Pandemic," *Foreign Affairs*, July/August 2005.

[82] Julie E. Fischer, *Stewardship or Censorship? Balancing Biosecurity, the Public's Health, and the Benefits of Scientific Openness* (Washington, DC: The Henry L. Stimson Center, 2006), p. 33.

[83] For an example of one effort to develop and harmonize international biosafety and biosecurity best practices, see the International Council for the Life Sciences, online at http://www.iclscharter.org/.

[84] For example see, Kansas NBAF Task Force, http://www.nbafinkansas.org/task_force/.

[85] Alternatives for Community & Environment, *What's Wrong With a BSL4 Bioterrorism Laboratory at Boston Medical Center?*, March 10, 2004, http://www.ace-ej.org/BiolabWeb/Biolabdocs/. What'swrongwithaBSL4labatBMCMay04.pdf.

[86] Ronald Atlas, "Toward Global Harmonization for Control of Dual-use Biothreat Agents," *Science and Public Policy*, vol. 35, no. 1 (February 2008), p. 26.

[87] Kathleen Vogel, "Framing Biosecurity: an Alternative to the Biotech Revolution Model?," *Science and Public Policy*, vol. 35, no. 1 (February 2008), p. 45-54.

[88] The American Society for Microbiology has advocated this path. See Jocelyn Kaiser, "Senate Bill Would Alter Biosafety, Select Agent Rules," *Science*, June 20, 2008.

[89] Executive Office of the President, The White House, *Executive Order 13486: Strengthening Laboratory Biosecurity in the United States*, 74 Fed. Reg. 2289-229 1, January 9, 2009.

[90] These individuals may appoint designees to represent them on the working group.

[91] Hugh Auchincloss, National Institute of Allergy and Infectious Diseases, National Institutes of Health, Testimony before the House Committee on Energy and Commerce, Subcommittee on Oversight and Investigations, October 4, 2007.

[92] Steven Kappes, U.S. Department of Agriculture, "Welcome and Opening Remarks," U.S. Government Trans-Federal Task Force on Optimizing Biosafety and Biocontainment Oversight Public Consultation, Washington, DC, December 8, 2008.

[93] Department of Health and Human Services and U.S. Department of Agriculture, *Charge, Trans-Federal Task Force on Optimizing Biosafety and Biocontainment Oversight*, http://www.hhs.gov/aspr/omsph/biosafety 120 1.pdf.

[94] Secretary of Health and Human Services, *Charter National Science Advisory Board for Biosecurity*, Department of Health and Human Services, Washington, DC, March 28, 2008.

[95] http://oba.od.nih.gov/biosecurity/biosecurity_documents.html.

[96] U.S. Government Accountability Office, *High-containment Biosafety Laboratories Preliminary Observations on the Oversight of the Proliferation of BSL-3 and BSL-4 Laboratories in the United States*, GAO-08-108T, October 4, 2007.

[97] For example, see Edward Hammond, Director of The Sunshine Project, before the House Committee on Energy and Commerce, Subcommittee on Oversight and Investigations, October 4, 2007; U.S. Government Accountability Office, *High-containment Biosafety Laboratories Preliminary Observations on the Oversight of the Proliferation of BSL-3 and BSL-4 Laboratories in the United States*, GAO-08-108T, October 4, 2007; and The Commission on the Prevention of WMD Proliferation and Terrorism, *World at Risk* (New York: Vintage Books, 2008).

[98] The survey commissioned by HHS exemplifies this issue. The response rate to the survey was under 50%. Constella Health Sciences, *Final Report: Survey for Determining the Location, Capacity, and Status of Existing and Operating BSL-3 Laboratory Facilities within the United States*, June 2, 2005.

[99] Letter from Representative John Dingell, Chairman of House Committee on Energy and Commerce, and Representative Bart Stupak, Chairman of the House Committee on Energy and Commerce, Subcommittee on Oversight and Investigations, to President Bush, August 8, 2008.

[100] For additional information on this facility and a more thorough discussion of its needs assessment, see CRS Report RL34 160, *The National Bio- and Agro-Defense Facility: Issues for Congress*, by Dana A. Shea, Jim Monke, and Frank Gottron.

[101] Such a capability was required through Executive Office of the President, The White House, *Biodefense for the 21st Century*, Homeland Security Presidential Directive (HSPD) 10, April 28, 2004. See also CRS Report RL32891, *The National Biodefense Analysis and Countermeasures Center: Issues for Congress*, by Dana A. Shea.

[102] For example, advisory panels and groups have recommended the expansion of large animal high-containment facilities for several years. See, for example, USDA Strategic Planning Task Force on Research Facilities, *Report of the Strategic Planning Task Force on USDA Research Facilities: Report and Recommendations*, August 1999.

[103] For example, see Jocelyn Kaiser, "Accidents Spur a Closer Look at Risks at Biodefense Labs," *Science*, vol. 317 (September 8, 2007), pp. 1852-1853.

[104] See CRS Report RL32391, *Small-scale Terrorist Attacks Using Chemical and Biological Agents: An Assessment Framework and Preliminary Comparisons*, by Dana A. Shea and Frank Gottron.

[105] Jocelyn Kaiser, "Senate Bill Would Alter Biosafety, Select Agent Rules," *Science*, June 20, 2008.

[106] For more information on the National Academies committee, see online at http://www8.nationalacademies.org/cp/ projectview.aspx?key=49063.

[107] Office of the Inspector General, *Inspection of the FBI's Security Risk Assessment Program for Individuals Requesting Access to Biological Agents and Toxins*, Evaluation and Inspections Report I-2005-003, March 2005.

[108] See discussion in the section "Select Agent Program" regarding disposal of pathogens rather than registering under the Select Agent Program.

[109] For information on Department of Defense biological surety programs, see Kathleen Carr, Erik A. Henchal, Catherine Wilhelmsen, Bridget Carr et al., "Implementation of Biosurety Systems in a Department of Defense Medical Research Laboratory," *Biosecurity and Bioterrorism: Biodefense Strategy, Practice, and Science*, Vol. 2, January 1, 2004, pp. 7-16; and Department of the Army, *Biological Surety*, Army Regulation 50-1, July 28, 2008.

[110] Department of Justice, *Transcript of Amerithrax Investigation Press Conference*, August 6, 2008, http://www.justice. See also Scott Shane, "Portrait Emerges of Anthrax Suspect's Troubled Life," *The New York Times*, January 3, 2009.

[111] 42 C.F.R. § 73.15.

[112] Gigi Kwik Gronvall, Joe Fitzgerald, Allison Chamberlain, et al., "High-Containment Biodefense Research Laboratories: Meeting Report and Center Recommendations," *Biosecurity and Bioterrorism: Biodefense Strategy, Practice, and Science*, vol. 5, no. 1 (2007), pp. 75-85.

[113] For a discussion of some of these reporting mechanisms, see CRS Report RL3 1983, *Health Care Quality: Improving Patient Safety by Promoting Medical Errors Reporting*, by C. Stephen Redhead.

[114] Accidents involving select agents are required to be reported to HHS or USDA under the Select Agent Program.

[115] Gigi Kwik Gronvall, Joe Fitzgerald, Allison Chamberlain, et al., "High-Containment Biodefense Research Laboratories: Meeting Report and Center Recommendations," *Biosecurity and Bioterrorism: Biodefense Strategy, Practice, and Science*, vol. 5, no. 1 (2007), pp. 75-85.

[116] Center for Biosecurity, University of Pittsburgh Medical Center, *Preventing and Deterring Biological Attacks: Priorities that Should Emerge from the WMD Commission Report*, December 19, 2008.

INDEX

D

E

F

G

H

I

J

L

M

N

O

P

Q

R

S

T

U